Survival in the Construction Business:

Checklists for Success

Thomas N. Frisby

Survival in the Construction Business:

Checklists for Success

Thomas N. Frisby

R.S. MEANS COMPANY, INC.
CONSTRUCTION CONSULTANTS & PUBLISHERS
100 Construction Plaza
P.O. Box 800
Kingston, MA 02364-0800
(617) 585-7880

This book was edited by Jeff Goldman and Mary Greene. Typesetting was supervised by Joan Marshman. The book and jacket were designed by Norman Forgit. Illustrations by Carl Linde.

Printed in the United States of America

10 9 8 7 6 5 4 3 2 1

Library of Congress Cataloging in Publication Data

ISBN 0-87629-153-1

Table of Contents

List of Checklists

Foreword

Almost every facet of the construction industry is undergoing some form of transition, as can be seen in buzz words like "Construction Management," "Fast Track," "Privatization," and "Partnering." Speed is also becoming a dominating factor, as there is less time for architects to design the structure (and consequently, less time to coordinate drawings, check dimensions, etc.) and the construction schedule is often tight and non-extendable.

Tools have been developed for managing projects at the speed required in this new construction era. The Critical Path Method (CPM) or other network analysis schedule devices are even specified in the contract documents. Millions are expended annually in generating these sophisticated computer-driven tools which, too often, are used ineffectively on the job, where they are not updated or made known to the subcontractors.

With the intense time restrictions, the frustration of incomplete drawings, and the absence or misuse of effective management tools (such as CPM), it is no wonder that job disputes, job losses, and claims sometimes seem to engulf our companies. Owners and architects, protecting themselves from lawsuits, are more often including "risk-transferring" clauses that attempt to shift the financial risk of designer- or owner-caused delays to the contractor. Owners are demanding higher standards in the field, and are increasing their requirements for contractors who seek to prove their costs for extra work or impact. Too often, the contractor's response is in the form of a lawsuit, with its attendant costs and effect on the firm's resources.

Indeed, claims and lawsuits may seem like an inevitable byproduct of the construction industry as it is structured today. At the same time, however, the transitional period that the construction industry is undergoing also presents an opportunity to reduce adversarial relations and upgrade management by all parties. It becomes clear that an improvement in the team effort of owner, architect, contractor, subcontractor, supplier, and workforce is in everyone's best interest—to build, cost, and schedule effective projects in accordance with the specified quality.

This book is written on the thesis that good old-fashioned fundamentals are the starting point, which everyone in the firm must not only understand, but also carry out. A general contractor who does not perform scheduling effectively cannot expect to be a successful construction manager. A contractor whose quality control program is lax cannot expect to be profitable (or even to survive) in the superfund-type projects which provide future opportunities. Though the industry is in transition, the successful contractor must completely master the basics before contemplating embarking on the road to change.

Acknowledgments

The construction industry is one of the most challenging of the truly free enterprises in America. It is also one of the highest risk, and potentially lowest profit industries. It has been my professional home for a third of a century, and has been very good to me. This book, aimed at improving the management practices of contractors, reducing their risks, and increasing their profits, is my own meager contribution to this great industry.

I would also like to acknowledge the inspiration—for this and other endeavors—which continues to come from Thelma Frisby and Nancy Butchko.

The book's readability is due to the editing of Mary Greene.

Chapter One

Managing Information with Checklists

Chapter One

Managing Information with Checklists

Many of the thousands of builders who go into business for themselves are able to stay in business only a few years before being overcome by the competition or their own inadequacies. The problem is that while there are many builders, there are few *contractors*. A *builder* is one whose company can place concrete, erect steel, pull wire or perform the functions required to install materials. A *contractor* must deal in contracts, and the work required to fulfill those agreements. A contractor must have expertise in construction, project management, contract management, and entrepreneurism, as well as marketing, technical capability, logistics management, decision-making management, information systems, financial management, legal management, and personnel management, as illustrated in Figure 1.1. A construction contractor must have many and more diverse skills than those required by almost any other industry.

Building construction is one of the highest risk industries. The major risks include:

- The contract
- The economy
- Weather, strikes
- Bid errors
- Differing site conditions
- Changes
- Impact of claims
- Productivity of personnel
- Availability of credible information
- Availability of competent personnel/laborers
- Potential bankruptcy of owner or subcontractors who owe you money/services
- The value system of others

Some of these risks are preventable; others may be mitigated. All must be recognized and managed.

This book is written for those who *manage*, whether they manage construction companies, the acquisition of projects (marketing and estimating), or the projects themselves. It is written for those who manage risks and make decisions that affect the businesses in the construction industry. Part I concentrates on the role of the individual who manages the construction firm. Part II addresses estimating and scheduling in the context of the firm's overall success. Part III covers crucial factors in project management and superintending (such as subcontractor management, handling changes and delays) that spell the difference between survival and disaster.

Information as a Basis of Management

The basis of successful management is information. In the construction industry, information is the lifeblood of all decision-making, whether it involves choosing a new market to enter, which job to bid or not to bid, how much of a contingency to include in the labor estimate for a risky job, or the recognition and processing of a change or claim under the contract. Information, of course, is crucial to *every one* of the many steps in the construction process. Just consider the steps that are involved in a "generic," fixed price construction contract, as shown in Figure 1.2.

Each of the "generic" steps shown in Figure 1.2 also has subparts. Consequently, there are a myriad of details that must be managed. Letting any of them slip can cause a contractor to be unsuccessful on a particular job, or even in his overall business. For example:

A contractor performing excavation encounters unexpected boulders. The few log borings included in the bid documents did not reveal subsurface boulders. However, the bid documents referenced a soils report in the architect's office. The contractor's estimator did not go to the architect's office to read the full soils report. If he had, he would have discovered that the presence of boulders at the building site could be anticipated based on previous work done in that area. The financial risk of encountering the subsurface boulders may very well belong to the contractor. He did not include this factor in his bid, and cannot now recover the

The Expertise Required of the Successful Contractor

- Building (technical expertise)
- Contract management
- Decision management
- Entrepreneurism
- Marketing
- Financial management
- Information systems
- Legal management
- Logistics management
- People management
- Forecasting
- Diplomacy
- Negotiation

Figure 1.1

Steps involved in a "Generic" Fixed-Price Construction Contract

☐ Owner's idea

☐ Market feasibility studies

☐ Site search and site feasibility study

☐ Scope development

☐ Conceptual estimate

☐ Economic estimate

☐ Preliminary design

☐ Factored estimate

☐ Project financing commitment and land acquisition

☐ Complete design

 ____ Site investigation

 ____ Drawings

 ____ Specifications

 ____ Constructability/value engineering analyses

 ____ Quality assurance

☐ A/E's detailed estimate

☐ Order long lead time materials and equipment

☐ Bid package preparation and pre-qualification of contractors

☐ Contractor's bid preparation

 ____ Site visit

 ____ Pre-Bid meeting

 ____ Subcontractor bidding/selection

 ____ Quantity takeoffs

 ____ Supplier quotes/commitments

 ____ Preliminary plan and schedule

 ____ Bid summary

☐ Post bid bond and tender bid

☐ Bid award/contract negotiations/payment scheduling

☐ Make financial security arrangements

 ____ Bonds (Performance, Payment, Subcontractor, etc.)

 ____ Insurance

 ____ Interim financing

Figure 1.2

Steps involved in a "Generic" Fixed-Price Construction Contract

☐ Secure building permits

☐ Purchase long lead time materials and equipment

☐ Award subcontracts/purchase agreements

    ____ Negotiate subcontracts/purchase orders

    ____ Receive, review, and process shop drawings/submittals

☐ Plan and schedule the project

☐ Set up project controls/organization

☐ Mobilize on the job site

    ____ Provide site access and parking/transportation arrangements

    ____ Provide job site security (guardhouse, fences, etc.)

    ____ Provide toilet facilities

    ____ Provide job office and supplies, warehouse, other temporary buildings

    ____ Provide temporary utilities

☐ Perform construction activities

    ____ Receive, store, and handle temporary and permanent materials/equipment

    ____ Provide and dispense tools and construction equipment

    ____ Provide temporary enclosures and safety equipment/features

☐ Prepare area for work

☐ Mobilize workers, equipment, and materials at task sites

☐ Supervise the work

    ____ Interpret drawing and specs

    ____ Develop short-term plans based on project schedule and progress

    ____ Develop safety program

    ____ Inform workers/subcontractors of task, method, location, and duration

    ____ Inspect work

    ____ Perform necessary quality control

    ____ On-site meetings

    ____ Log worker/equipment time and materials used

    ____ Owner-initiated change orders

    ____ Quality, subcontractor-initiated changes

    ____ Evaluate performance

Figure 1.2 (continued)

Steps involved in a "Generic" Fixed-Price Construction Contract

    ____ Work with municipality inspections, etc.

☐ Execute the activities and concurrent tasks

☐ Owner project meetings

☐ Perform cost accounting and control operations

    ____ Assemble cost information

    ____ Evaluate project completion status/subcontract status

    ____ Prepare invoices and pay bills

    ____ Summarize information and compare to plan/bid expectations

    ____ Prepare report based on summary

    ____ Analyze methods for improvements

    ____ Update schedule/revise plans

    ____ Demobilize workers and equipment

    ____ Cleanup area and dispose of waste materials

    ____ Remove, clean and store reusable (temporary) materials and tools

☐ Prepare and execute punch list

☐ Concur that project is substantially complete

☐ Demobilize from job site

    ____ Remove all temporary buildings

    ____ Remove fences and temporary utilities

    ____ Clean and prepare site for finish site work

    ____ Complete site work

☐ Provide owner with all warranties, extra materials, as-built drawings, equipment manuals, key, etc.

☐ Occupancy inspections with owner, municipality

☐ Owner move-in

☐ Facility start-up

☐ Guarantee of construction/workmanship

☐ Call-backs

Figure 1.2 (continued)

cost under **changed conditions** because the information was, in fact, provided by the owner. The contractor simply neglected to obtain it.

> A contractor decides to bid on work in a booming construction market in another state. He applies his standard unit prices to the job **without** checking the availability of labor in that area. It turns out that labor is scarce and productivity is low. Failure to check out that one fact—the supply of competent labor—could lead to a project without a profit. In fact, the job could very well lose money.

Every step in the construction process is important, and perhaps vital. Proper and thorough information is directly linked to a project's success. Again, the lifeblood of any management endeavor is *information*.

Accurate information enables managers to make sound decisions.

Complete information permits managers to make decisions without fear of being "blind-sided," or surprised later on.

Concise information enables managers to assimilate a large amount of data.

Timely information eliminates indecisiveness and provides necessary momentum.

The Use of Checklists

How can managers ensure that adequate attention is being given to each step in the construction process? It may be that no system can be devised to guarantee that absolutely nothing is missed and no detail overlooked. However, there is a system that aids a manager and his organization in always remembering the key information and risks involved in each step of the construction process. This system is *construction checklists*.

A pilot of a 747 may have taken off and landed safely hundreds of times, yet he still follows multiple checklists to ensure that none of the crucial details are overlooked. Managers in construction must also have a system to remember the crucial details.

The checklists the pilot uses do not fly the plane—they simply remind the pilot of what he must do in order to fly it. The same is true of the construction checklists which are presented in the following chapters. They will not manage a company, estimate a job, or build a project. They do serve as reminders to the manager to develop the most reliable, timely, and complete information possible with which to manage—profitably—the risks inherent in a construction company or a construction project. The checklists should help answer the following questions:

- What does a construction company president do to manage a company profitably?
- What does a chief estimator do to manage the estimating process?
- How does a project manager manage?

The checklists provided in this book are grouped into three main categories based on the diverse goals of a construction company. Chapters 1 and 2 contain the checklists used to manage the overall construction firm. Chapters 3 and 4 provide the checklists needed to manage the risks inherent in acquiring and estimating work. Chapters 5 through 8 describe how to manage the project successfully using checklists. Each section includes a description of the risks and steps in the management process, followed by checklists used to gather critical data for managing these steps and risks.

These checklists have been proven in the industry, in use by successful contractors. If the checklist system is to be effective, the following points must be observed.

- The checklists must be used routinely.
- The information provided through the checklists must be concise and relevant to the management or decision-making issue.
- Someone in the organization must be held accountable for the accuracy and completeness of the information recorded on the checklists. In the case of the 747, a mechanic must sign off on the repairs he has made and the equipment he has checked. It should be the same with data presented in the construction company.
- Checklists should be tailored to *your* company.
- *All* management personnel should be taught how to use checklists properly.
- The checklist should never be used as a substitute for the human mind and its ingenuity—only as an aid in the exercise of that ingenuity in decision-making and managing.

Chapter Two

Organizing, Marketing, and Managing Your Company

Chapter Two

Organizing, Marketing, and Managing Your Company

The first thing a construction company president must do is to "get his act together," or organize the business to run smoothly and profitably. He or she must then keep his/her act together as the market, economic conditions, and internal factors change, as inevitably they will.

A construction company president, then, must manage:

- The internal organization (Chapter 2)
- The acquisition of work (Chapters 3 and 4)
- The performance of work (Chapters 5, 7, and 8)
- Financial accounting of the results (Chapter 2)

These elements must be managed in the context of:

- The market (Chapter 2)
- The changing economy (Chapter 2)
- Ever-evolving construction trends (Chapter 2)

To effectively manage each of these vital areas, the president must be aware of the risks inherent in them. He must also have accurate and timely information from competent, qualified personnel (who are accountable for that information) in order to be able to make effective and profit-oriented decisions regarding those risks. There is a major distinction between taking calculated risks and gambling: the gambler rolls the dice without studying the odds or being able to change them, whereas those who take calculated risks are aware of the risks as well as what can be done to manage them. They look at the information critically and analytically, taking nothing for granted and testing what others say against their own experience and knowledge.

This chapter will explore some of the major risks that may arise in running a construction company and coping with changing market and economic conditions. Having identified the risks, we will then provide some tools—in the form of the following checklists—for working with and limiting those risks.

1. "What Is the Market?" (Figure 2.2)—This checklist is used to analyze the market that is appropriate and available to the construction firm.

2. "Determining our Present Capabilities" (Figure 2.4)—This checklist is geared to determining the company's capabilities, including any special strengths or weak points.
3. "The Market Plan" (Figure 2.5)—This checklist is used to develop the firm's own market plan and goals.
4. "Requirements of PPOIC" (Figure 2.16)—serves as a guide for managing **P**eople, **P**lanning, **O**rganization, **I**mplementation, and **C**ontrol.
5. "Performance Expectation" (Figure 2.17)—provides reasonable expectations that the president of a construction firm may have regarding the performance of foremen, superintendents, and project managers.

All of these checklists are designed to address the most important, though often overlooked factors in organizing, marketing, and managing a construction firm.

Managing the Market

The "market" refers to the opportunities that exist to successfully bid and build projects. In a free market system, supply and demand dictate the number of opportunities that exist and where they exist. Supply and demand also sets the level of competition. Competition can limit the market for some firms whose unit prices are higher or safety record poorer (resulting in higher overhead) than others.

When we say "manage the market," we mean two things:
- Know what the opportunities are in terms of both geographic locations and types of work
- Know how to organize the company to take advantage of those opportunities. This is what we call "entrepreneurism."

Entrepreneurism

Entrepreneurism is at the very heart of the free enterprise system, yet it can be a difficult term to define and perhaps an even more difficult concept to follow successfully. In this book, we will treat entrepreneurism as a *process*, not a fixed state. It is a process by which a company:
- *Identifies* the needs of the marketplace.
- *Develops* such that it can fulfill those needs. It must develop in a way that:
  — *costs less* than the competition can do it
  — makes a *profit* for the company
  — *accepts rather than defies risk*

Thus, entrepreneurism is not flamboyance, dice throwing, or magic. It is, in fact, *fundamental* to profitable management, and is the ultimate responsibility of everyone in the construction company, under the direction of the company president.

Entrepreneurism is a living and dynamic process, not a legend carved into a granite tablet. It involves:
1. Defining the market
2. Organizing to meet the needs of the market
3. Obtaining work in that market
4. Performing the work profitably while satisfying the customer
5. Accounting for finances
6. Developing personnel

Entrepreneurism also involves *intelligent, deliberate change and improvement*. This means developing a solid organization and being constantly aware of conditions that require changes (either fine tuning or dramatic) in order to remain an effective and profitable competitor. The organization that steadfastly refuses to change or that changes for the sake of change is usually called an *ex-contractor*.

How to Identify the Profitable Markets

Successful marketing equation:
Market needs = Company's capability to perform that kind of work profitably.

A company must avoid imposing its own needs or identity on the marketplace. For example, the company that is greed-oriented (a profit at any price) is often eventually a victim of its own short-sightedness. A company that remains in a dying market or fails to keep abreast of modern management systems is looking at a potential decline in profits. Thus, the firm must look at the *needs* of the market, *where* those needs exist, and *who* has those needs.

Business Cycles

All needs of the market are cyclical. A particular need (such as rental housing or petrochemical plants) will sometimes be unsatisfied in a given location. That is, people need more of these facilities than are available. Generally, when the need becomes known, contractors strive to quickly establish themselves as builders of the kinds of structures that are in demand, whether houses, condos, offices, or industrial space. Before long, the intensity of the need is reduced (that is, there are a sufficient number of facilities) because contractors flocked to a booming, or "hot" market.

The first step in marketing is to determine not only the needs of the marketplace, but also the *constraints* involved in fulfilling those needs. For example, there may be a great need for housing, but high interest rates, recession, or other factors may reduce the *actual need* to an *effective need*. An effective need is one that can be fulfilled profitably, under existing economic conditions.

The contractor must determine the *point in the market* at which he can expect to earn the greatest profit. The graph in Figure 2.1 shows that in some circumstances, a booming market may produce busted contractors because of the great competition for competent personnel and supervisors, material, and other resources. For example, during a construction "boom" in Calgary, Canada in the 1970's, many contractors sought a "piece of the pie." Several won large contracts, only to lose money because they could not obtain a skilled workforce or adequate supplies in competition with all the other work going on.

There are two very dangerous times to enter a market. The first is when the market is at its lowest point, when contractors will bid at a very low margin to get a job in order to maintain their cash flow. The second dangerous time is the market peak where contractors are coming in from outside the market area to take advantage of the demand. The result is a shortage of craftsmen, supervisory personnel, and materials, and consequently, higher prices. Knowing this, the construction company president should take the following actions.

In an upswing market, he should search the market for qualified supervisory and craft personnel *before* signing contracts for the work. He should also work with the accountant, surety, and banker to ensure that the appropriate cash flow, surety credit, and management tools are in place to handle growth. The contractor should further survey the labor market, and seek out qualified subcontractors and suppliers.

In a declining market, the contractor should be alert and ready to reduce overhead, while surveying other markets—all *before* the plunge. In this diverse country, it seems that there is always a pretty good market *somewhere*. The contractor should also be careful to avoid the burden of costly equipment ownership when the market is at its lowest point.

The construction company president must anticipate market conditions and figure out how to continue competing profitably. As an example, simply reading the newspaper tells you that prison construction, airport expansion, and toxic waste and acid rain abatement projects are coming markets. One can also surmise that the building market is slowing down, simply by checking out vacancy rates published monthly.

Most presidents of construction firms can estimate, run a job, and read a financial report. They do reasonably well managing their companies in stable markets. However, too few are skilled at marketing. As a result, the companies over which they preside generally do not fare well in periods of either major market expansion or contraction.

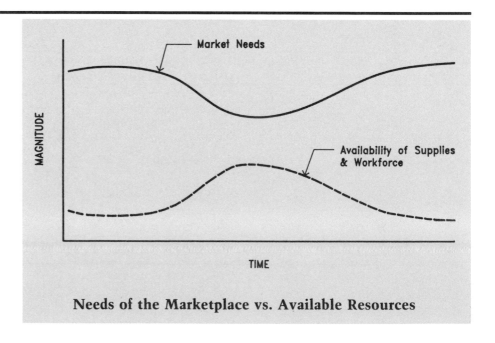

Needs of the Marketplace vs. Available Resources

Figure 2.1

Another risk of a boom market (which is generally inflationary) is that the contractor's business tends to grow too quickly, with inadequate organization to manage his larger company effectively. Furthermore, having acquired a large, fixed overhead for the larger market, he may be scarcely able to handle the bills when the bottom inevitably drops out of the cyclical market.

Defining the market extends beyond determining the "hot" areas where the demand exceeds the supply, to finding out *business volume*. This kind of information can be obtained from many sources including, but not limited to:

- *Dodge Reports*
- Utility company reports (telephone company, electric company)
- Building permit trends
- HUD Reports
- Government agency (federal, state and municipal) capital spending plans
- Financial institution reports and forecasts
- *Kiplinger Reports*
- *Engineering News Record*
- Local trade groups (Associated General Contractors, Associated Builders and Contractors, Chamber of Commerce, etc.)

Contractors, large or small, must be aware of changing economic forces. For example, in defining the current market, the construction contractor might consider toxic waste management sites, a demand for which exists across the United States, and will take years to be filled. In terms of the company's organization, trends such as Construction Management may be considered. "Partnering" (which is a joint venture between the owner and the construction team) and privitization (which is the public sector turning some of its functions, such as prison construction and operation, over to the private sector), are other coming trends that some firms should explore. New methods and materials will also have an effect on the construction industry and should be considered. For example, super-conductivity is being diversified. (This process uses porcelain to transmit electricity with minimal loss of power.)

Top managers must be aware of economic cycles, and must position themselves to be on the *leading edge* of change, for if they wait until the market peaks, they are just another contractor in the pack.

Where Is the Market?

About fifteen years ago, most contractors did most of their work *within* twenty-five miles of their office. Now, most contractors do most of their work more than twenty-five miles *away* from their office. The market for building construction varies throughout the U.S. In the seventies and early eighties, Texas was a "hot" market, but the market declined along with the oil industry. Now, cities like Philadelphia, Baltimore, and Boston, pronounced dead in the seventies, are flourishing. In fact, there seems to be a trend toward a bi-coastal nation where most of the population and wealth will be concentrated on the coasts. The point is that all markets are like the shifting sands of a desert. No one geographic area remains "hot," and most do not remain "dead." The astute construction contractor remains sensitive to these shifting economic trends, anticipating both positive and negative changes. That same astute contractor will also recognize that doing work outside of the firm's market area can be a great risk and, therefore, demands a cautious and prudent approach. In fact, a hallmark characteristic of the successful contractor is *cautious change*.

Who Is the Market?

Within each generic market (housing, petrochemical, government, etc.), it is necessary to define the "Who." That is, "Who is the Owner?" "Who needs the facility?" "Who will be commissioning substantial work?" The contractor must develop a list of the "Who's" and actively seek to be put on their bidders' lists or to be properly pre-qualified for their work. In addition, the contractor should learn as much as possible about the characteristics of the companies that need the facilities.

For example, if the "Who" is the Corps of Engineers (COE), the contractor must learn what it is like to do business with the COE, its standards of acceptance, its paperwork requirements, and the individuals in charge of facility construction. If the "Who" is a private developer, the contractor should investigate its financial integrity, its payment and change order history, and who it has done business with in the past (and why).

In addition, the contractor must ask "What does this party require of a contractor?" If the "Who" is the COE, it wants a bonded contractor with the *lowest* bid. However, the COE may have additional requirements, such as contractors who can design and build, as well as a number of minority contractors, and/or women in construction.

A private owner may want something other than the lowest bidder. A particular party may require or emphasize:

- A construction manager
- Unusual quality
- Proven integrity
- A non-claims-oriented contractor
- Other more subjective criteria, in addition to the price, such as on-time performance and responsiveness

A market profile checklist can be a helpful tool. An example is shown in Figure 2.2. The rough outline for the Market Profile checklist can be developed by the construction firm's president, and organized and updated monthly by his/her secretary. The sources of information for the monthly updates are shown in the checklist, but the secretary should also clip pertinent articles from the local newspapers and trade journals. Thus on a monthly basis, the president has his/her own version of the "Kiplinger report," which reflects what is happening or may be happening in the company's market area.

From this brief checklist, the president of the company can get a "feel" for market trends and the factors that affect those trends. The checklist can be used to develop an excellent list of the people and organizations that have the resources—and the need—to build, and even many of the specific projects that may be

What is Our Market?

This checklist should be prepared annually and updated every six months or more often.

1. What is the **size** of our market? (in terms of geographic area and category of work, such as housing, pollution control, sewer, highway, etc.)

   _____

   $ _____ volume last year

   $ _____ expected volume this year

   _____ upswing

   _____ downswing

   _____ trend

   Source of information: _____ (such as Dodge Reports)

   _____

   What is the size of other markets our company may be interested in penetrating? Describe market area we may be interested in.

   _____

   _____

   _____

   $ _____ volume last year

   $ _____ expected volume this year

   _____ upswing _____ downswing _____ trend

2. What **spending factors** seem to be affecting your market? (use up or down arrow to indicate the trend)

   _____ Federal Spending

   _____ State Spending

   _____ Industrial Development

   _____ Commercial Development

   _____ Foreign Investment

   _____ Municipal Spending

   (Source of information: Write your congressman for federal data; your state senator for state data; the telephone company, utility company, Chamber of Commerce for non-government spending; the local school board, Director of Public Works, and Sewage and Water Districts for local government spending.)

Figure 2.2

3. What additional **competition** is coming into the area?

Company	Size	Market it is Pursuing
_____	_____	_____
_____	_____	_____
_____	_____	_____

(Source of data: Local trade associations, surety company, unions)

4. What are the **economic trends** in our market area? (Provide percentages.)

Inflation _____

Unemployment _____

Prime Rate _____

Office Bldg. Occupancy Data _____

(Source: Local bank, Chamber of Commerce)

5. What are some of the major projects coming up this year?

A. Government

Federal	**State**	**Local Government**
☐ GSA	☐ Highway	☐ Pollution Control
☐ USBR	☐ College	☐ Utility (sewer lines, etc.)
☐ COE	☐ Building	☐ Schools
☐ Navy	☐ Renovation	☐ Public Building
☐ NASA		
☐ DOE		
☐ DA		

(Source: All government projects are funded by appropriation or bond issue, which is public knowledge.)

B. Quasi-Public Agencies

☐ Electric Company

☐ Gas Company

☐ Telephone Company

(Source: These companies have capital expenditure plans which may be obtained by simply requesting the information.)

Figure 2.2 (continued)

What Is Our Market?

C. Private Sector

☐ Office Buildings

☐ Bank Buildings

☐ Housing Developments

☐ Shopping Centers

☐ Warehousing

(Source: Dodge Reports, of course. However, personal rapport with lending institutions, architects, engineers, and suppliers will often provide very useful information.)

Figure 2.2 (continued)

forthcoming. The president can analyze these trends to determine which direction his company should take.

For example, if the private sector market is beginning to stagnate, the company president may consider other options, such as military or municipal construction. If he decides to penetrate one of these markets, he knows in advance that his company had better learn how to do business with the government. Paying attention (in advance) to the trends not only helps the company take advantage of lucrative opportunities; it also saves it from sinking like the Titanic as many did in Dallas and Houston (as well as other areas whose economies collapsed) during the mid-eighties, by moving toward better markets before it is too late.

Getting an Edge

Once the construction firm president knows what the market is and has analyzed trends as to where it is going, the next step is to organize in such a way that the company can compete profitably. As stated previously, management has three primary functions:

- Selling its services
- Performing its services
- Financially accounting for what it has sold and produced

Selling One's Services

Selling one's services, more commonly termed, *marketing*, is a very broad function. Marketing for a successful contracting firm is more than classy brochures, mailers, entertainment, and door-knocking. Rather, it requires a commitment to a certain concept of marketing, which affects every fiber of the company. Marketing is essentially creating something unique about your company that the market needs now. It is *your* advantage.

Differential advantage! Edge! These are what separate you from the pack. What do you have to offer that is an *advantage* to the client, that is better or different from what is offered by the rest of the approximately 300,000 contractors out there? The unfortunate mentality of the construction industry seems to be that the differential advantage, or *edge*, is simply having the lowest bid. The fact is that the *differential advantage*, the underlying concept of effective marketing, must embody every system, every function, every resource that a contractor has. Differential advantage must be planned! And that planning must be executed! Thus, *planning* is the essence and underpinning of productivity.

The president of a construction firm begins this planning with a market plan or a business development plan. In other words, the president must:

- Plan to get the work
- Plan to manage the work
- Plan to develop the best possible team of personnel
- Plan to beat the competition
- Plan to stay in business
- Plan for his/her successor

This plan should be the result of the collective input of all company managerial personnel. It should be based on "hard" data, and not just a "wish list." It should provide management with a vehicle for continuous appraisal of the company's strengths, a frank and candid statement of the company's shortfalls, and a blueprint for improving itself before proceeding into new arenas of unknown or unperceived risks. See Figure 2.3 for an illustration of the plan's goals.

Using the Self Analysis Checklist in Figure 2.4 will help to determine the areas in which the contractor is "in the pack" with his competitors, and those in which he stands out with a unique service or advantage. To the extent he is "in the pack," he really has nothing to sell, that is, it is difficult to market convincingly his company's services to an owner if he cannot answer these owner questions: "Why should we contract with you? What do you have to offer that distinguishes you from your competitors?"

Only when the contractor can answer: "*These* are my unique abilities, my differential advantages," can he be continuously effective as a viable contractor with an organization honed to achieve maximum productivity, job after job.

The Self Analysis Checklist also helps to determine the company's weak areas. It encourages constant vigilance so that the firm's strengths can be protected, and its weaknesses corrected. It also requires that one have an organized source of information to maintain that vigilance on a routine basis.

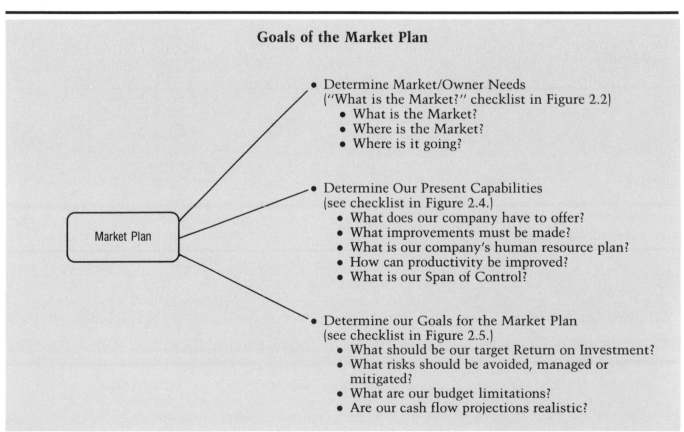

Goals of the Market Plan

- Determine Market/Owner Needs
  ("What is the Market?" checklist in Figure 2.2)
    - What is the Market?
    - Where is the Market?
    - Where is it going?

- Determine Our Present Capabilities
  (see checklist in Figure 2.4.)
    - What does our company have to offer?
    - What improvements must be made?
    - What is our company's human resource plan?
    - How can productivity be improved?
    - What is our Span of Control?

- Determine our Goals for the Market Plan
  (see checklist in Figure 2.5.)
    - What should be our target Return on Investment?
    - What risks should be avoided, managed or mitigated?
    - What are our budget limitations?
    - Are our cash flow projections realistic?

Market Plan

Figure 2.3

Determining Our Present Capabilities (and Unique Strengths)

The following is a checklist for performing your own self-evaluation.

1. FINANCIAL RATIOS

	Avg.	Our Company
Current Ratio (current assets minus current liabilities)		
Current Debt to Net Worth		
Net Fixed Assets to Net Worth		
Pre-tax Net Income to Net Worth		
Gross Billings to Net Worth		
Gross Billings to Working Capital		
Pre-tax Net Income (Loss) to Gross Billings		
Pre-tax Net Income to Total Assets		
Aging – Receivables		
Aging – Payables		
(Three Year Summary)		

2. FINANCIAL STATUS

Working Capital (current assets minus current liabilities)

(Adequate) _____   (Inadequate) _____

If working capital is inadequate, why?

        Job Losses   _____

        Growth   _____

Financial Planning Tools

	In Place	Adequate	Inadequate
Budget			
Market Plan			
Cash Flow Projection			
Adequate Labor Cost Reporting			

Figure 2.4

Determining Our Present Capabilities

Labor History

% of Jobs Over Labor Budget   _____ %

% of Jobs Equal to or Less than Labor Budget   _____ %

$ Overrun (Underrun) Before Budget Last Year $ _____

Cash Flow Management

Excellent _____ Good _____ Fair _____ Bad _____

Interest

Amount Paid $ _____

As % of Gross Profit   _____ %

Billings

Present under (over) Billing $ _____

Percentage of Market   _____ %

3. PERSONNEL

Successor Plan   Yes _____ No _____

Training Program   Yes _____ No _____

Financial   _____ _____

Contract Mgt.   _____ _____

Technical   _____ _____

Foreman   _____ _____

Craftsmen   _____ _____

Turnover of Key Personnel _____ (No.)

Capability of Supervisory Personnel
(Rate as Excellent, Good, Fair, Unsatisfactory)

Name	Present	Potential
President		

Figure 2.4 (continued)

Determining Our Present Capabilities

Profit Producers

(List from foreman up to the key profit producers in the company)

Name   Contribution

_____   _____

_____   _____

_____   _____

_____   _____

Profit Eaters

List all supervisory personnel from the foreman up who were responsible for lost money.

Name   Amount   Reason

_____ $ _____ _____

_____   _____ _____

_____   _____ _____

_____   _____ _____

Bonus/Profit Sharing Plan Yes _____ No _____

Profit Before Plan $ _____

Profit After Plan   _____

Safety Record (bonus)

Comparison with Industry Average   _____ %

4. CORPORATE MANAGEMENT TOOLS   Yes   No

Bid/No Bid Decision Process   _____ _____

Corporate Objectives & Market Plan   _____ _____

Excellent Relationship with Surety   _____ _____

Excellent Relationship with Bank   _____ _____

Human Resources Plan   _____ _____

Management-Oriented   _____ _____

Committed to Planning   _____ _____

Productivity Program   _____ _____

Figure 2.4 (continued)

Determining Our Present Capabilities

5. ESTIMATING

Assessment of Workload	Good _____	Bad _____	
Win/Loss Ratio	_____ %		
Adequacy of Pre-Bid Site Evaluation:	Excellent _____	Good _____	Fair _____

Does Estimating Consider:

	Yes	No
Task Analysis	_____	_____
Pricing Scheduling	_____	_____
Manloading Curve	_____	_____
Cash Investment	_____	_____
Cash Flow	_____	_____
Return on Investment	_____	_____
Risks (Quantification)	_____	_____
Special Requirements	_____	_____
Adequate Documentation	_____	_____
Terms & Conditions	_____	_____
Labor Market	_____	_____
"Differences" from Typical Job	_____	_____

6. JOB SET-UP

Do job planners consider:

	Yes	No
Schedule	_____	_____
Adequate Schedule	_____	_____
Good Statement of Work (Subcontractors)	_____	_____
Priorities Established	_____	_____
Proper Manning	_____	_____
Job Budget	_____	_____
Task Analysis	_____	_____
Risk Identification	_____	_____
Documentation Requirements	_____	_____
Method of Managing Interfaces	_____	_____
Previous Experiences Considered	_____	_____

Figure 2.4 (continued)

7. PERFORMANCE

Corrective & Warranty Work $ _____   _____ %

Quality Control Program   Yes _____ No _____

Safety Program   Yes _____ No _____

Productivity Program   Yes _____ No _____

On-the-Job Training   Yes _____ No _____

Cost Control Program   Yes _____ No _____

Schedule Updating   Yes _____ No _____

Turnover   Yes _____ No _____

    Project Manager _____

    Superintendent _____

    Foremen _____

    Craftsmen _____

Where do you usually do best
(e.g., concrete)? _____

Where do you usually lose
your money (e.g., concrete)? _____

**Commitment to Contract
Compliance:**   Yes _____ No _____

    (Proactive)   (Reactive)

Close-out Problems

Review by Job

8. CONTRACT MANAGEMENT

Is job function properly understood by:

President   Yes _____ No _____

Project Manager   Yes _____ No _____

Superintendent   Yes _____ No _____

Foreman   Yes _____ No _____

Are these procedures properly understood:

Notifications   Yes   No

Documentation   Yes _____ No _____

Figure 2.4 (continued)

Determining Our Present Capabilities

Acceleration	Yes _____	No _____	
Constructive Changes	Yes _____	No _____	
Impact	Yes _____	No _____	

9. STRENGTHS AND WEAKNESSES

From the self-evaluation analysis, we can then develop a list of our strengths and our weaknesses. For example, a contractor should consider the following categories (these are illustrative and are not intended to be complete):

Category	Strength	Weakness
A. Market Planning	_____	_____
B. Estimating		
Conceptual estimating	_____	_____
Capacity to develop value engineering cost proposals in pre-construction	_____	_____
Adequacy of historical pricing data	_____	_____
Relationship with subcontractors	_____	_____
Relationship with suppliers	_____	_____
Experienced estimators in:		
concrete	_____	_____
structural	_____	_____
process	_____	_____
mechanical/electrical	_____	_____
commercial	_____	_____
industrial	_____	_____
office	_____	_____
low-rise	_____	_____
high-rise	_____	_____
renovation	_____	_____
heavy	_____	_____
other	_____	_____
Adequate field input	_____	_____
Adequate procedures	_____	_____
Adequate bid documentation	_____	_____

Figure 2.4 (continued)

Determining Our Present Capabilities

Category	Strength	Weakness
Cash flow	_____	_____
Technical risks	_____	_____
Operational risks	_____	_____
Differences	_____	_____
Schedule and production flow		_____
Success ratio in jobs by size:		
Under $1,000,000	_____	_____
$1,000,000 – 5,000,000	_____	_____
Over $5,000,000	_____	_____
Success ratio in jobs by function:		
Free-standing structures	_____	_____
Water treatment	_____	_____
Multi-family	_____	_____
Institutions	_____	_____
Municipal	_____	_____
Federal	_____	_____
Engineering	_____	_____
Other		
commercial	_____	_____
industrial	_____	_____
heavy	_____	_____
office/low-rise	_____	_____
office/hi-rise	_____	_____
renovation	_____	_____
Size and organization of staff	_____	_____

Figure 2.4 (continued)

Determining Our Present Capabilities

Category	Strength	Weakness
C. **Project Management**		
Experienced and Successful Managers in:		
Types of work		
Free- standing structures	_____	_____
Water treatment	_____	_____
Multi-family	_____	_____
Institutions	_____	_____
Municipal	_____	_____
Federal	_____	_____
Engineering	_____	_____
Commercial	_____	_____
Industrial	_____	_____
Heavy	_____	_____
Office/low-rise	_____	_____
Office/hi-rise	_____	_____
Rchab. waste	_____	_____
Residential/development	_____	_____
Size of jobs		
Under $500,000 – $1,000,000	_____	_____
$1,000,000 – $5,000,000	_____	_____
over $5,000,000	_____	_____
Bid work	_____	_____
Negotiated work	_____	_____
Construction management	_____	_____
Experienced superintendents in:		
Types of work		
Free-standing structures	_____	_____
Water treatment	_____	_____

Figure 2.4 (continued)

Determining Our Present Capabilities

Category	Strength	Weakness
Multi-family	_____	_____
Institutions	_____	_____
Municipal	_____	_____
Federal	_____	_____
Engineering	_____	_____
Commercial	_____	_____
Industrial	_____	_____
Heavy	_____	_____
Office/low-rise	_____	_____
Office/high-rise	_____	_____
Rehab. work	_____	_____
Residential/development	_____	_____
Size of jobs		
Under $500,000 - 1,000,000	_____	_____
$1,000,000 – $5,000,000	_____	_____
Over $5,000,000	_____	_____
Bid work	_____	_____
Negotiated work	_____	_____
Construction management	_____	_____
Quality of foreman	_____	_____
Scheduling	_____	_____

Figure 2.4 (continued)

Determining Our Present Capabilities

Category	Strength	Weakness
Performing own work in:		
Earthwork	_____	_____
Concrete	_____	_____
Structure	_____	_____
Process Piping	_____	_____
Carpentry	_____	_____
Other		
_____	_____	_____
_____	_____	_____
Finish work		
_____	_____	_____
_____	_____	_____
Adequate cost control	_____	_____
Effective forecasting	_____	_____
Effective subcontractor management	_____	_____
Effective logistics management	_____	_____
Effective contract management	_____	_____
Effectiveness in problem prevention	_____	_____
Effectiveness in recovery plans	_____	_____
Effectiveness in working with owner	_____	_____
Effectiveness in working with architect	_____	_____
Documentation	_____	_____
Job close-out	_____	_____
Productivity	_____	_____
Working with unions	_____	_____

Figure 2.4 (continued)

Determining Our Present Capabilities

Category	Strength	Weakness
Financial Management		
Market Plan – bid/no bid criteria	_____	_____
Budget	_____	_____
Cash flow projections	_____	_____
Cash flow management	_____	_____
Job cost systems	_____	_____
Profit/Loss and balance sheets	_____	_____

Figure 2.4 (continued)

Having identified the company's weaknesses, the president can then develop a plan for overcoming them. Attention should also be given to providing the services required in the marketplace by a particular client.

Figure 2.5 is an outline of a Market Plan. The Market Plan becomes the navigational aid the contractor uses to stay on the right course. Having studied the market and evaluated its strengths and weaknesses, the contractor should know—within reasonable parameters—the markets in which he/she can make a profit, and the markets that should be either avoided or entered with great caution. With the Market Plan, the contractor charts the right course, avoids dangerous territory, and periodically corrects and refines his course. The Market Plan establishes the boundaries of risk beyond which the contractor should not go, and the revenue and profit goals he wishes to achieve.

The Market Plan should be referred to during every "Bid/No Bid" decision meeting as both the criteria for acquiring projects and a score card ("How are we doing and what, if anything, do we need to improve on?").

Failure to plan and manage one's position in the marketplace is perhaps best illustrated by the disastrous growth attempts made by many contractors. Growth is more often the force behind the failure of construction companies than any other single factor. Yet it is a time-honored belief that a company *must* grow. "You can't stand still, you know, or you'll shrivel up," is a classic opinion held by construction company presidents.

Management of Growth

Figure 2.6 shows "Frisby's Hypothesis": most construction companies that experience exceptional growth within a short time frame will either go bankrupt (the "X" curve) or *almost* go out of business before regaining control and developing a more moderate and long-term growth pattern (the "Y" slope).

To understand the theory that exceptional growth often leads to a financial crisis, it is necessary to examine the way in which construction companies are created. For the most part, when a construction company is being formed, its founder develops a Span of Control. By *Span of Control*, we mean that the founder went through the mental exercise illustrated in Figure 2.7. From this simplified list of questions, the founder can now quantify and qualify his effective Span of Control, as demonstrated by Figure 2.8.

Figure 2.8 shows the limits that a contractor establishes for his operation, saying to himself, "As long as I operate in the market that I know, with experienced and reliable people, and do not push my cash position to the breaking point, I can operate profitably and effectively. I am, therefore, going to stay inside these limits; *my growth shall never exceed my capacity to manage my business!*" Generally speaking, so long as the contractor operates within this span of control, his profit and revenue growth can progress along an ever-ascending curve.

However, many, if not most, contractors grow beyond their capacity to manage their business and find themselves with a breach in their Span of Control. (This situation is illustrated in

The Market Plan

Based on the "Determining Our Capabilities Checklist," now develop your market plan and goals. (Note: This form has been filled in with sample information to illustrate the use of the checklist.)

I. Span of Control	A. Job Type/Complexity	*Shopping centers, warehouses. Avoid Government market until we have developed capabilities as set forth below.*
	B. Location	*Within Union 501's jurisdiction.*
	C. Degree of Risk	*No job which exceeds three times our net worth.*
	D. Characteristics	*Beware of unreasonable or financially-weak owner.*
	E. Architect/ Engineer Characteristics	*Beware of architect with reputation for poor plans, poor decision-making.*
	F. Labor/ Equipment Supply	*Be careful of markets where labor/ equipment is in very short supply — where we can get the job but can't build it productively because of work force shortage.*
	G. Cash Flow Required/Expected	*Do not bid on job where cash flow requirements exceed present capability. Seek jobs which permit payment for mobilization, have an average of 5% retainage; do not have a propensity for dragging out close-out and final payment.*
	H. Gross Profit Opportunity	6-10%
	I. Competition	*We will avoid banging our heads against the wall in the bid commercial market where the competition is so high and the return so low.*
	J. Present Workload	*Beware of overloading estimating. Do not hire project manager or superintendent "off the street."*
	K. Uniqueness of Project, OR Experience in this Type or Location of Job	*If we have no experience in this type project, we should generally avoid bidding it unless it can be joint ventured with an experienced contractor or we can get proven supervision we know.*
	L. Surety Limits	*$25,000,000 backlog and $5,000,000 single-job limit. We must prove to our surety that these limits are too restrictive and are limiting our opportunities.*
	M. Job Mix	Length of Jobs *Strive for a mix of long-term and shorter turnaround jobs for improved cash flow.* Degree of Subcontracting *Use only bonded subs. This year try to minimize risk and cash flow by subbing out all high risk work.* Level of Risk *Risk will be within acceptable limits. Risk will be identified and properly priced or managed.*

Figure 2.5

The Market Plan

II. Financial Goals		
	Return on Investment	*14%*
	Net Profit	*2.8%*
	Gross Revenue	*22,500,000*
	Liquidity & Cash Flow Goals	
	Receivable Aging	*30 days*
	Payable Aging	*35 days*
	Current Ratio	*2:1*
	Working Capital (current assets - current liabilities)	*$1,500,000*

Clean-up of In-process Work

Retainages:

Jobs	$ Outstanding	Comments
Insurance Office Building	*$100,000*	*Get curtain wall sub to repair leaks.*
Health Care Center	*28,000*	
City Office Building	*78,500*	*Get HVAC sub to balance system. Repair asphalt paving where compaction was out of spec.*

Cost and Overhead Reduction Goals

Item	$	Date
Equipment Sale	*$150,000*	*3-18*

Identify the Specific Steps that must be accomplished to attain the foregoing goals:

1. New Markets:
   A. *Study how to do business with COE and GSA.*
   B. *Begin looking for Project Manager with Federal experience.*
   C. *Consider bidding small, low-risk job with subcontractor who is heavy in Government work.*

2. Existing Personnel:
   A. *All supervisory personnel are weak in contract administration. Schedule in-house program with emphasis on Government contracts.*
   B. *Estimating staff is not fully capable of using computer-assisted programs. Must continue to develop this capability. Ratio of 12% wins too low. Must analyze where we are missing jobs.*

3. Additional Personnel:
   A. *Begin looking for Project Manager with Government experience.*
   B. *Begin looking for two superintendents with hospital experience.*

4. Capital:
   Surety wants us to get rid of some rolling stock and real estate to add to liquidity of firm.

5. Equipment:
   Need an effective preventive maintenance program. Productivity on scrapers is seriously down because of poor repair and lack of maintenance.

6. Education/Training:
   We must have more one-on-ones with foremen to teach them how to enhance productivity and to improve quality. Safety record must be improved to be competitive.

7. Productivity:
   Must Improve in: Equipment maintenance; safety; material handling; and in the pre-planning of jobs.

8. Other:
   A. **Joint Ventures** - *Consider doing J.V. with a large mechanical sub on waste water treatment plant.*
   B. **Mergers/Acquisitions** - *We will begin looking for a small process piping company to acquire.*
   C. **Add In-house Capabilities** - *Personal Computers to all Project Managers. But we must use them effectively.*

Figure 2.5 (continued)

Figure 2.9.) They have penetrated their effective Span of Control by *reacting* to market conditions rather than *pro-actively continuing to plan* their destiny. In almost every case, when growth threatens a contractor's effective Span of Control, profits decline and financial disaster may follow. This turn of events is usually preceded by "bad" jobs which fail to meet their productivity and financial goals.

How often is it discovered that a company suffering financial malaise has:

- Taken on a much larger project than it is accustomed to?
- Hired untested personnel who do not have a track record within that company?
- Entered a new market (such as a commercial contractor entering the plant or industrial field, or a "bid contractor" entering the design/build field, etc.)?
- Penetrated a new geographic area without adequate planning or staff?
- Stretched the limits of its institutional credit (surety and bank including insurance and bonding capacities)?
- Failed to provide development opportunities for its personnel?
- Inundated itself with assets (plant and equipment) which deprived it of flexibility when the market was down?

These (and other) problems are signs of management outgrowing its capacity to run its business. The problem behind these unfortunate developments is simply that the contractor failed to plan for and manage growth. Contractors realize that their

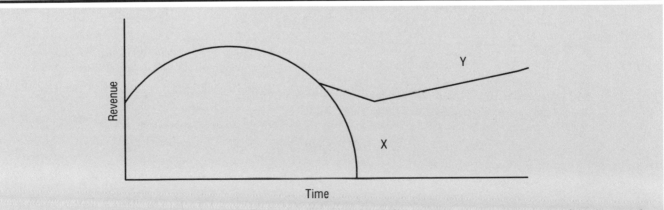

"Frisby's Hypothesis:" Simply stated, most construction companies that experience exceptional growth within a short time frame will either go bankrupt (the "x" curve) or *almost* go out of business before regaining control and developing a more moderate and long-term growth pattern (the "Y" slope).

Figure 2.6

estimates must be properly managed, projects must be managed, and people must be managed. But they fail to consider that *growth must be managed as well*. The following example is a case in point.

> *Dinosaur Constructors, Inc. (a mythical contractor based on a composite of actual circumstances) was a ten-year-old general contracting firm building projects valued at about $55,000,000 annually. With business conditions slightly down in its own state, Dinosaur feverishly began to seek out opportunities elsewhere. The path taken by this company is shown in Figure 2.10. Is it surprising that this was the company's last project?*

What Is My Span of Control?

1. How much cash can I raise?
2. What personnel can I hire with confidence?
3. What owners have I worked with who thought well of me?
4. What architects have I worked with who thought well of me?
5. What kind of construction do I do best?
6. What kind of tools and equipment will I need to get started? Can I lease it?
7. Which surety will I work with? What kind of information will it require in order to provide credit?
8. Which bank will I work with? What kind of information will it require in order to provide credit?
9. What are my weaknesses?
10. What can I do as well or better than others? (What are my strengths?)
11. Where is the market? How competitive is it?
12. What kind of financial controls do I need?

Figure 2.7

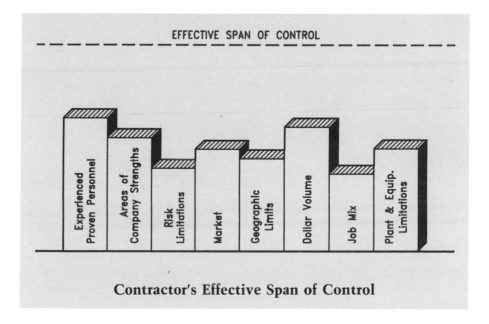

Contractor's Effective Span of Control

Figure 2.8

We see that growth is not only revenue growth, but expansion into a new territory, a new market area, or a new way of doing business. It is growth by merger or acquisition, for instance: a general contractor taking on a process piping division, or a mechanical contractor bidding as a general, or an electrical contractor entering the design/build field, or doing business with the government for the first time . . . and so on.

Step 1 Lag

Each of these growth areas *must* be managed. To assist contractors with this task, we have developed the concept of Step 1 Lag. By *Step 1 Lag*, we mean simply this: Before you grow, you must plan for that growth in Step 1. Growth follows, or lags behind, the Step 1 planning process; thus, Step 1 *Lag* (see Figure 2.11). This is a continuous, dynamic, and evolutionary process aimed at developing and using the contractor's muscle in its business opportunities, rather than stumbling because of its Achilles Heel.

At the core of Step 1 Lag is the intelligent development of a market plan, carefully conceived and monitored, re-evaluated and updated during the year. The following section describes the tools a construction firm needs to orient itself toward successful growth management.

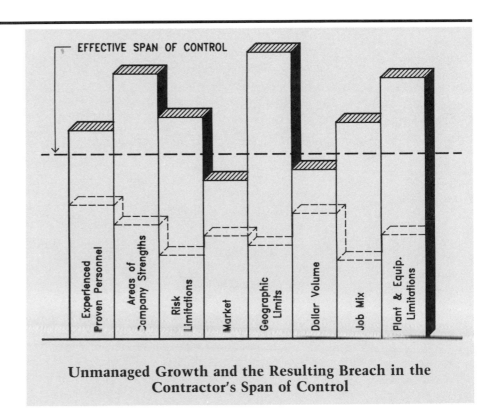

Unmanaged Growth and the Resulting Breach in the Contractor's Span of Control

Figure 2.9

Organizational Structure and Goals

Construction firms employ a number of different organizational structures to operate their businesses. Often these structures are depicted by neat blocks and lines that are more meaningful to the draftsmen who made the chart than to the people in the company. It is not the organizational chart that is important, but the *concept* behind the business. As Peter Drucker observed, *"organization follows concept."*

An organizational structure that can sustain economic growth incorporates the elements of both a democracy and a dictatorship. Key company personnel participate in the development of company goals, which is a democratic process. On the other hand, the construction firm president must act alone in making final decisions and standing by them; thus the dictatorship. The goal is to meld together the company's resources in order to maximize total company productivity. This is accomplished through a *goals program* (the bottom line goal being *Return On Investment*, or *ROI*). In this goals program, all supervisory personnel can *participate* in establishing successful growth. They are *responsible* for attaining this goal, and are also *rewarded* for their success. The following is a four-step process for the successful implementation of this organizational concept.

Faulty Growth Pattern of Sample Contracting Firm	
New Action	**Prior Track Record**
1. Negotiated...	1. Had always previously bid its work.
2. ...$15,000,000...	2. Largest previous project was $7,500,000.
3. ...fast-track, construction management type contract...	3. Had always been the general contractor.
4. ...for the construction of concrete floors for a huge assembly building	4. Had performed this type of work on many occasions.
5. ...in a location 1,000 miles from home	5. Had never operated in this state before.
6. ...as an open shop contractor...	6. Had always been union. Tried "double-breasted" operation for the first time in this project.
7. ...with project manager whose experience was in office buildings.	7. Generally developed its project managers from the ranks.

Figure 2.10

Step 1, illustrated in Figure 2.12, amounts to the president or chief executive officer of the construction company coming up with a basic market plan and company goals, using the tools and checklists previously described.

In Step 2 (illustrated in Figure 2.13), the company president presents the preliminary market plan and goals to the operating department heads. The department heads are asked for responses to these goals based on the evaluations of questions listed underneath each department. The responses and evaluations are then submitted to the president. A dialogue is established between the president and the department heads. Initially, this is carried out on a one-on-one basis, and finally, a meeting is held with all of the department heads.

At this point, all of the key personnel have had an opportunity to participate in the setting of goals. In Step 3, the president sees that all of the department heads *commit* to the attainment of these goals. Step 3 is shown in an organizational chart form in Figure 2.14. This figure shows that the firm's revenue, overhead, budget, and return on investment are the primary determinant in the company's goals (along with input from the surety). The market plan is developed, and the president communicates these priorities to the various departments within the firm.

The company now has both a destination and a road map for getting there. The key people who can make it happen have helped to develop the plans and have committed to making those plans become a reality. The president has allowed democracy to work (i.e., participation in the process), but alone is the one who must make the final decisions and continue to insist on the efforts required to achieve the goals set forth. This is the

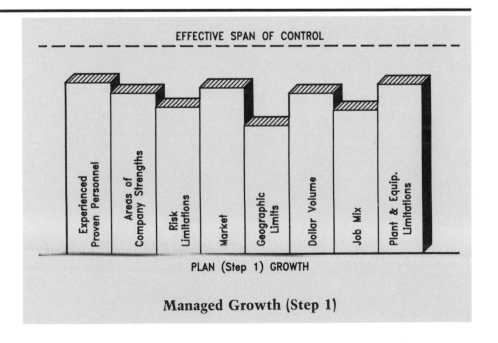

Managed Growth (Step 1)

Figure 2.11

Basic Market Plan and Company Goals

Step 1: Drawing up a Basic Market Plan and Company Goals (checklist for the construction company president)

- ☐ What is the market?
- ☐ Where is the market?
- ☐ Where is the market going?
- ☐ What does our company have to offer?
- ☐ What improvements need to be made within our firm?
- ☐ What is our strategy for market entry?
- ☐ What should our goals be?
  - ____ Return on investment
  - ____ Budget
  - ____ Revenue

Figure 2.12

"dictatorship" element—another way of saying that a construction company must have a very strong, highly-qualified chief executive officer who encourages participation by his or her subordinates.

Note that the accounting department does not prepare the overall budget, but rather supplies the input. Too often, the planning function is turned over completely to accounting. Then, without in-depth knowledge of the company's operating strengths, weaknesses and potential, or of the outside market, the accounting department simply extrapolates from last year's results. This approach can make it difficult to obtain the necessary commitment to the firm's goals from the other personnel—because it is the *accountant's* plan, and not their own.

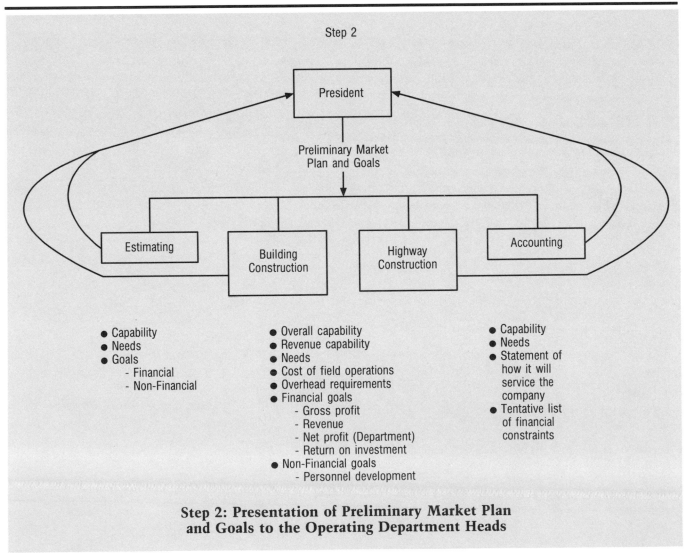

Step 2: Presentation of Preliminary Market Plan and Goals to the Operating Department Heads

Figure 2.13

Step 4 is an extension of Step 3. The department heads use a procedure similar to that previously used by the president to pass the company goals down to all managerial personnel. The department heads then set goals for specific projects that are in line with the company's overall goals (see Figure 2.15).

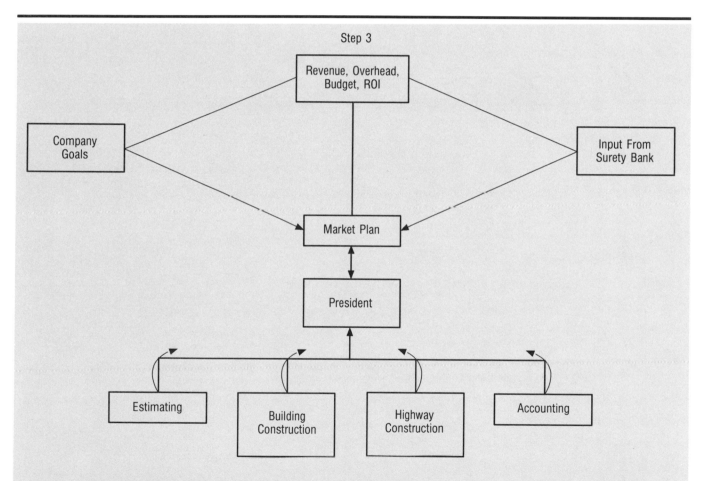

Step 3: Obtaining the Commitment of Department Heads to Company Goals

This figure indicates that the development of the market plan is not the individual work of any one person, including the president. The president integrates the information concerning strengths, weaknesses, and goals, both from key company personnel and from outside resources (such as the bonding company and the surety). From this information, a market plan is developed, from which realistic financial targets can be created in the form of revenue and profit projections.

Figure 2.14

The company goals can be "democratically" developed under the "dictatorship" of the president, using these four steps. The desired end result, as stated early in this section, is to increase total company productivity. But just what is a company's total productivity, and how is it measured?

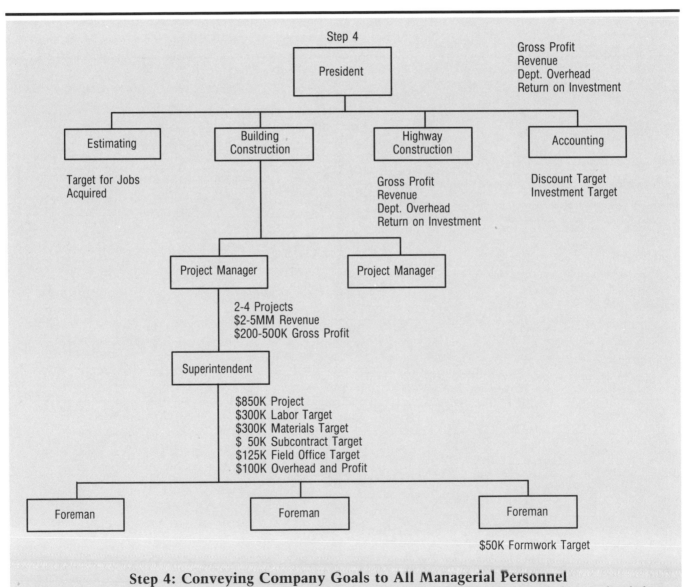

Step 4: Conveying Company Goals to All Managerial Personnel
This matrix indicates that every supervisory level is essentially a *profit center* and that each of these profit centers must know their goals and commit to them.

Figure 2.15

Productivity

Productivity (or lack of it) is spoken about often in the construction industry, and is seen as an area in crisis. Yet somehow, no one ever does anything about it. This situation has led to reactive management myths about both the causes and solutions to productivity problems. One of the greatest obstacles to improving productivity is the myth that "productivity means working harder." That simply is not true. Productivity can be improved by working *smarter*. For example, a major area of labor law involves what happens to equipment between its arrival on the job and its installation. Using the planning checklists in Chapter 5 will cause supervisory personnel to "think out" in advance the most productive flow of equipment once it arrives on the job site.

Another reactive myth is that the answer to increased productivity lies exclusively in the field operation—"those crews have to work more effectively." Certainly the field operations are central to optimum production. However, the greatest crisis in construction industry productivity is in *management*, which fails to educate field personnel in planning, scheduling, and contract management. This problem can, in part, be overcome by ensuring that the basic guidelines set forth in Chapters 5–8 are followed.

It is sometimes popular to blame eroding productivity on unions, the government, or a faltering work ethic. An element of truth can be found in any of these allegations, but generally, the president of a successful construction company accepts responsibility for the "bottom line," and does not rationalize or blame others.

Return on Investment

Construction management today is beginning to measure the successful operation of a construction company based upon the *Return on Investment After Tax* (ROI). To compute the ROI, one simply uses the formula:

$$\text{(Net Profit After Tax)} - \text{(Net Worth)} = \text{ROI}$$
$$\text{(or Annual Investment)}$$

By using ROI as a measurement of productivity, one can identify a whole range of management functions that affect a construction company's overall productivity. "Company productivity" can be scrutinized instead of telescoping in on field productivity alone. For example, a company's cash management concept and performance may have as great an influence on the bottom line as meeting (or exceeding) the labor limitations in the budget. The decision to bid the right jobs and avoid the bad jobs is as important, difficult, and influential to the bottom line as any other management decision or activity in a company. The field force can be working at maximum efficiency, but if your company has bid the financially weak owner or the big job in another state outside your technical expertise, it is likely that the job will be a loser and your ROI will suffer as a result.

Proactive Management

Proactive Management is "telescoping management," that is, looking forward at the priorities and potential problems that lie ahead. It is identifying those potential priorities and problems, then assigning responsibility to the proper person to ensure that they are taken care of *before* they become a reality. Reactive management is waiting until there is a crisis and then doing something about it.

Reactive management must become *proactive* management. The management of a firm must accept its responsibility for productivity, and refrain from laying the blame on such factors as unions, government over-regulation, and lost pride in workmanship. Each of these excuses does indeed limit increased production, and later chapters of this book will provide recommendations for dealing with each of them. However, it is the central theme of this book that *management must face up to its responsibility for productivity*, and by doing so, it can achieve results that substantially improve profits, open up markets, and reduce the risks that prevail in our industry.

Productivity can be thought of in parallel terms to the land management program of the farmer. To have a "bumper crop" in construction, to achieve the most fruits of our labor at the best prices and highest profit, contractors must take a systems approach to their own companies and to the construction industry itself. Contractors must *optimize* each function, each activity, and, above all, each decision, the way a farmer must optimize the use of each acre. Contractors must *harmonize* their market plans with their capacity to produce, their home office functions with the field force, subcontractor with general contractor, contractor with designer, supplier with constructor, union with management, government with free enterprise, profit with incentive, merit with reward. In order to squeeze out every dollar from every element of its operation, a contractor must bid the right jobs, buy or lease the right equipment, and use the right people in the right places.

General contractors and construction managers must themselves be permitted to do the scheduling job they were hired to do. They must integrate the supply industry into the construction industry. They must stop treating subcontractors like second-class citizens. All of this is necessary if field productivity is to be improved on their projects, and claims avoided.

Contractors must rid themselves of the major problems that interfere with productivity, such as unnecessarily protracted delays by architects in issuing change orders or approving shop drawings, inadequate soils investigations, unnecessary accelerations (usually constructive accelerations), poor drawings, lack of planning, unqualified subcontractors, and unrealistic expectations by the architect or owner.

A contractor is not just a constructor; he is a manager of construction, contracts, marketing, people, logistics, legal issues, money, and business. If a contractor is not all of these, and more, he is not an effective manager of productivity. The checklists in the following chapters provide the blueprints for creative thinking, memory jogging, and routine discipline to deal with these issues.

PPOIC: the Elements of Management

The proactive president, who manages company productivity, causes *all* operations of the company to perform as expected by paying close attention to the basic elements of management. We refer to these combined elements as *PPOIC*. They are:

- People
- Planning
- Organization
- Implementation
- Control

Figure 2.16 illustrates these elements, and the basic requirements for each.

It is up to the president of the construction firm to tie the entire organization together into a successful functioning package through People-Planning-Organization-Implementation-Control. Once the president has a market plan, he must ensure that the firm is organized properly to acquire work, produce it profitably, and manage the risks successfully. Each department head in the organization, and each person in each department, must know what is expected of him, and what he can expect of others above, below, and parallel to him in the company organization. This doctrine of expectation—of communicating what each person is to do as well as what each person can rely upon others to do—is the central nervous system of a good organization.

Unfortunately, these principles are rarely understood or implemented in construction companies. The performance expectation chart in Figure 2.17 can be helpful in making this analysis. It lists the major activities required for PPOIC management—from the foreman up to the top of an organization. It is no accident that *People* are at the top of the list in PPOIC.

The Requirements of PPOIC
(People, Planning, Organization, Implementation, Control)

The construction company president must ensure at every level of the organization that:

People...	• hired are qualified. • are trained for present and future jobs.	• are motivated. • are adequately rewarded.
Planning is occurring for...	• markets. • organization.	• job acquisition. • financial management.
Organization is continuing and active for...	• equipment and tools. • work packages to be done. • communication with all parties.	• material handling. • dealing with any crisis.
Plans are properly implemented...	• i.e., that people carry out that which was planned and organized.	
There is proper control...	• which requires a **system** to make sure that people carry out that which was planned and organized.	

Figure 2.16

While the products of the construction industry are *things*, those things are needed, designed, built, and paid for by people.

Figure 2.18 is another type of performance evaluation chart which shows how the people in the organization can be evaluated against the elements of PPOIC. When the evaluator, or manager, sees an area of weakness, he can then work with the individual to improve that area. More importantly, when *trends* of weakness (such as planning or contract management) show up in several employees, it is time for in-house training programs.

Financial Management

Financial management and productivity are blood relatives. That is, financial management involves not only a system for reporting what a company earns, but also a means for helping to earn it. In other words, financial management is "part of the action," a vital element in the company's productivity.

The best tool so far devised for measuring a company's productivity is Return On Investment, already mentioned as *Net Profit After Tax/Net Worth*. Another way of understanding ROI is:

$$\frac{\text{What I made \& could keep this year}}{\text{What I started the year with}}$$

Then, the ratio of what a company started the year with to what it could keep is its *Return On Investment*. The more productive a company is, the greater its earnings should be and the higher the ROI. The belief that ROI is governed only by good bidding and good field production is a fallacy! The management of finances may have as great an impact on ROI as the management of jobs in the field. Figure 2.19 illustrates some of the factors *outside* field production which influence earnings. While each of the items in Figure 2.19 are being intensely managed to increase earnings, the contractor must at the same time *protect* the investment by giving careful attention to the items shown in Figure 2.20. Of course, the elements affected—the company investment—should be monitored monthly as a continual scorecard of how well the firm is doing.

The intent of this section is not to provide a treatise on financial management, but rather to stress that:

- Financial management is not just an accountant's function.
- Financial management is the responsibility of everyone in the company.
- Through effective financial management, earnings can be increased while the investment is protected.

Performance Expectation for the Foreman

Required Skills and Responsibilities:

People...	• Make sure they know that they work for your company first, not the union. • Know their strengths and weaknesses. • Give clear instructions to the work force. • Train the work force.	• Show them how they fit into the schedule as a whole. • Set goals for the work force. • Insist on performance. • Work against absenteeism and drug/alcohol abuse.
Planning...	• Read applicable specs. • Review drawings. • Know required quality (working tolerances). • Know crew needed and output. • Consider requirements for an adequate work environment (e.g., controlling mud or working around it). • Know tool and equipment needs. • Create a material handling plan. • Know how much time it takes to complete a task.	• Know when each craft must have completed their work, so that subsequent work can be performed. • Know what is unusual about particular tasks. • Consider how to do this job better, to meet the goals of budget, quality, and time. • Consider the safety aspects of your methods. • Work with project management to develop overall project plan.
Organize...	• Inform crew of plan. • Inform others (suppliers, superintendent, inspectors) of plan in a timely manner.	• Ensure the availability of the right quantity and quality of tools, equipment, and material. • Plan the work environment for safety.
Implement...	• Carry out to completion what was planned and organized. • Make sure cost coding reporting is accurate, including: — documenting the progress of your work crews. — documenting changed work. — documenting crew movement, other effects on the crew.	• Know the details of the job. • Check the dimensions upon completion of the work. • Keep superintendent informed of progress or divergences from schedule. • Make sure preceding work is done; check dimensions, etc., before actually beginning work.
Control...	• Keep track of planned durations. • Keep track of quality. • Watch out for interferences and inform superintendent.	• Keep an eye on man-hours/production goals. • Listen to feedback/suggestions of crew. • Keep track of actual durations for use on next project with similar characteristics.

Figure 2.17

Performance Expectation for the Superintendent

Required Skills and Responsibilities:

People...	• Assist the foremen as follows: — training them to plan and set goals. — making sure that they have read the plans and specs and know the quality requirements. — showing them how they fit into the schedule as a whole. — encouraging creativity. • Maintain a positive working relationship with foremen and project manager.
Plan...	• Know the plans and specs intimately. • Establish a schedule; review, revise, and update it. • Establish short-term goals. • Advise staff of any problems in a timely manner. • Let others know what they need to do in a timely manner. • Cause subcontractors to plan. (Conduct a planning session.) • Work with foremen to ensure that each task is defined, planned, and executed. • Develop manpower curves. • Make sure quality standards are understood (including subcontractors). • Establish tool and equipment needs. • Communicate documentation requirements to foremen. • Know applicable general/special conditions requirements. • Consider how to meet indicated profit/schedule/quality for a successful job. • Work with engineering staff to define material priorities and required delivery dates. • Establish indirect (or non-critical) material requirements and delivery dates. • Work with project manager to establish the overall project plan and schedule.
Organize...	• Meet with foremen to discuss plans. • Make sure all others (suppliers/subcontractors/inspectors/other divisions) know what is expected, and when. • Ensure that proper materials/equipment are available or expedited as needed. Keep the work force in the field working with the required tools, equipment, and materials — on time, and performing the work properly. • Make sure that material handling is conducted most efficiently. • Establish administrative areas of responsibility with the project manager.
Implement...	• Carry out and cause others to carry out what was planned and organized. • Make sure cost reporting is accurate, including: — reliable cost codes, — adequate contractual documentation. • Know the details of the job.
Control...	• Keep track of the budget. • Keep track of quality. • Keep track of and document interferences. • Continually update the schedule. • Let others (foremen, project managers, subcontractors) know when their performance is good or inadequate. • Evaluate staffing. • Use manpower studies, productivity curves to your advantage.

Figure 2.17 (continued)

Performance Expectation for Project Management

Required Skills and Responsibilities:

People...	• Work with the superintendents; establish a positive working relationship. • Assist the project administrative engineering staff, including: — helping train them to plan, set goals. — making sure they have read plans/specs, know project quality requirements. — encouraging creativity. • Make sure that all project personnel know company policies. • Establish contact with personnel on the owner's staff who have control/authority on the project.
Plan...	• Work with the superintendents to establish a job plan. • Know the contract documents. • Know and manage the risks and differences of this particular project. • Act as decision manager. • Develop a complete schedule with others. • Know how the estimating department priced the job. • Establish short-term goals. • Establish a control plan for quality/submittals/documents. • Establish a material management plan. • Establish a contract management plan (include changes). • Work with superintendent, others to keep labor costs on or under budget. • Establish a meaningful cost control program. (Inform cost engineers of desired analyses.) • Establish what kind of installation/coordination drawings you require. • Establish a plan for continuous cost/schedule information. • Plan project close-out and turnover. • Make effective use of your computer(s). • Establish interdivisional relationships. • Make sure everyone understands his/her job. • Be sure you understand the function of the general contractor on each project. • Determine how to work effectively with the owner/architect. • Plan how to meet the budget.
Organize...	• Meet with superintendents, project engineers, schedulers, cost engineers, and other divisions to discuss plans. • Get input from each of these groups/individuals to make sure that they have planned effectively. • Meet with owner/architect to establish and review priorities. • Provide adequate, organized information for decision-making.
Implement...	• Carry out and motivate others to carry out what was planned. • Be intolerant of failures to follow the plan. • Know where you are at all times in relation to the schedule. • Ensure that project payments are collected and invoices paid.
Control...	• Keep track of the budget. • Keep track of milestones and other early warning signals; report to division management as required. • Update the schedule. • Let others (owner, architect, other divisions, subcontractors, suppliers) know when they are not doing their jobs. • Set the pace on the project; use the schedule as a motivational tool. • Review all documentation; if it does not reflect planned conditions, determine the differences, seek solutions, and plan and implement corrective strategies.

Figure 2.17 (continued)

Performance Evaluation

Position: Project Manager/General Superintendents
Dept./Div: Building

Employee _____ Supervisor/Reviewer _____ Date _____

Factors	–	=	+	N	Explanation of Rating: (Complete only if negative or exceptional)
PEOPLE					
Relationship with owners, architects, consultants and staff					
Working relationship with subcontractors					
Relationship with support personnel in main office					
Relationship with other divisions					
Work with estimators for proper turnover of project to operation.					
Develop subordinates.					
Assist project staff to accomplish project goals.					
Marketing our company					
Manage manpower loading at start, during, and at close of project.					
Relationship with trade unions					
PLAN					
Approach to project and discussions with project staff					
Strategy with project staff to meet project requirements					
Workable schedule with project staff; compare with estimators					

(–) Needs improvement to meet the job requirement.
(=) Fully meets job requirement.
(+) Exceeds job requirement.
(N) Check the "N" box if you have no opinion or category is not applicable.

Figure 2.18

Performance Evaluation

Factors	−	=	+	N	Explanation of Rating: (Complete only if negative or exceptional)
Change order procedure					
Safety program for the project					
Weekly meetings with owners, architects and project staff					
Quality control: foundation, concrete, rebar and finish work. Set goals for project staff, including hourly staff					
Productivity goals					
Project close-out (punch list, shop drawings, as built, warranty, etc.)					
ORGANIZE					
Equipment, materials and tools					
Subcontractors meeting					
Safety committee					
Pre-Con meeting with subcontractors and vendors (what we expect from them: quality, management of materials and manpower, safety)					
Quality control person/persons					
IMPLEMENT					
Execute change orders immediately.					
Safety program					
Quality control plan					
Productivity plan and goals					

Figure 2.18 (continued)

Performance Evaluation

Factors	–	=	+	N	Explanation of Rating: (Complete only if negative or exceptional)
Up-front pricing of change orders					
Project close-out					
CONTROL					
Financial status of project (pay request, project GOP, receivables, accounts payable)					
Report early warning signals to division manager immediately.					
Productivity review weekly with project staff					
Owner and architect					
Subcontractors manpower					
Material delivery for subs and our company					
Meet weekly with quality control persons. Respond immediately to problems.					
Create project schedule and update project schedule weekly. Reprint monthly if necessary.					
Review project strategy on weekly basis.					
Understand contract, specifications and drawings.					

Comments:

_____

_____

Improvements/deficiencies as compared to previous analysis

_____

_____
Signature of Reviewer

Figure 2.18 (continued)

Consider the areas in which money can be made on cash flow alone:

- Discounts
- Short-term investment of *float* (the period between deposits in the bank and the time that checks written on the account clear the bank)
- Short-term investment of *cash in bank*
- Managing the *receivables* intensely to keep the aging as low as possible, while pushing out p*ayables* as far as possible, without jeopardizing credit or discounts.

These techniques, and more, improve earnings and the overall Return On Investment. In fact, the finance department of the construction company should consider itself a *profit center*. There are companies that bill (invoice) $10,000,000 in sales annually and keep some portion of their receipts invested in short-term investments (such as T-Bills) for 5-10 days every 30 days. Even if the return is only 6%, the companies are improving their profit annually from the investment of their cash flows. Considering that the average construction company earns a total of only 1-2% net against revenue, such a money management program could have a substantial influence on earnings.

The concept of *staying power* is monumentally important to contractors. As a contractor stretches its capital to the limit, it reduces its capabilities to ride out a bad job, to "hang in there"

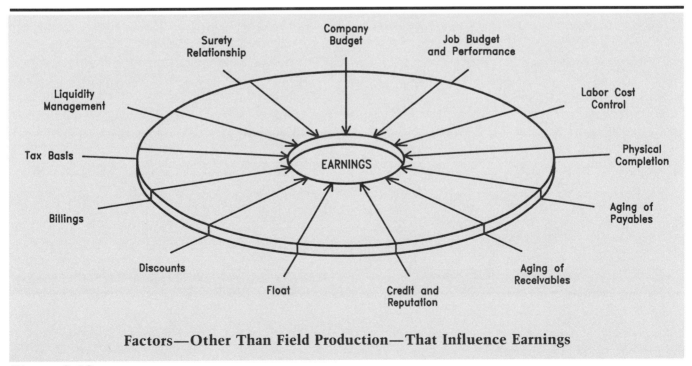

Factors—Other Than Field Production—That Influence Earnings

Figure 2.19

until it can get full recovery on a valid claim, and to not panic during the low points in the construction market. No goal is more important to a contractor's planning than to constantly develop adequate liquidity and to operate within it. In this way, the company is assured of having a the adequate staying power to survive the "lows" that are inevitable in this industry!

The Keys to Success

Extensive research has been conducted into the reasons behind the success and/or failure of contractors. The fact is, most contractors who go into business fail; some manage to stay in business for a long time, but never become a great success; and only a few thrive over the years. Figure 2.21 is an illustration of the characteristics of the successful contractor. This summary is based on research and personal observation. Every construction company president should analyze each of these characteristics to see how his or her approach compares with the model.

A proactive, not reactive, style of management is promoted by these management goals. Proactive management is constantly risk-proofing its operation by maintaining awareness of and control over potential hazards. Reactive management, on the other hand, is always rushing in to "put out fires." While one can obtain insurance against damaging forces outside one's control, it

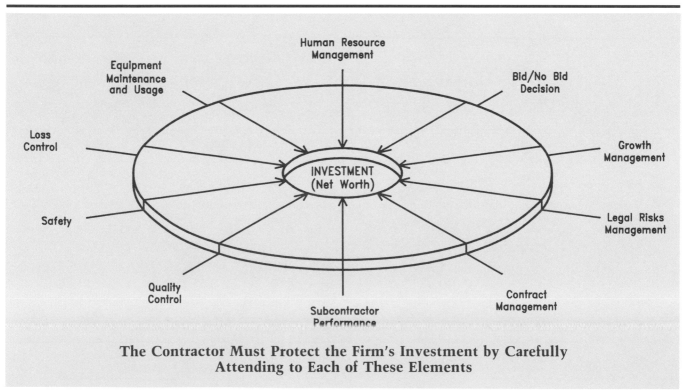

The Contractor Must Protect the Firm's Investment by Carefully Attending to Each of These Elements

Figure 2.20

Management Characteristics of the Successful Contractor

Management by the "Old Man"

All continuously successful construction companies have an "old man" (the president or CEO) who:
- Knows the business inside-out.
- Lets others know what he expects of them and demands performance.
- Instills a value system in employees and their work.
- Acts as a "guru," i.e., teaches others by his own standards, example, and coaching.
- Views intelligent change as a companion, not an adversary.
- Does not tolerate disloyalty, but does accept (however reluctantly) constructive debate until a decision is made.
- Does not cheat or tolerate cheaters.
- Knows the company's risks and how to manage them. As importantly, he makes sure that company personnel know the risks and how to manage them.
- Uses profit to build his company, not just his summer home.
- Never uses the scapegoat approach to management.

All "losers" have an "old man" who does not measure up to all of the above. If a company is having problems, and the president is considering firing someone, the first person who should be considered is the president.

People Management

- The construction industry is an industry of *people*.
- Greatness in this industry (regardless of all the advantages provided by computers and the latest technology) is still generated by *people*.
- The management of people separates contractors from ex-contractors, winners from losers.
- Management of people involves:
  - making a commitment
  - knowing how to carry it out
  - selecting people
  - training people
  - motivating people
  - rewarding people
- Finally, the buck stops with the person in charge — the president or CEO of the company.

Management of Priorities

- Establish priorities at the beginning of each job:
  - yours
  - the owner's
  - the architect's
  - subcontractors'
  - others
- Establish priorities thereafter on a weekly basis.
- Monitor them to ensure that they are attained.
- One of the priorities is to establish priorities.
- Establish priorities which prevent crises. A crisis is not a priority item, it is a panic situation.

Management of Continual Improvement

- Make a commitment to improvement.
- Know what needs to be improved.
- Know who needs improvement.
- Know when improvement is needed.
- Tie improvements to profit.
- Carry out improvement.
- Foster dedication to improvement in others.

Management to Avoid Repeated Mistakes

- The Frisby Learning Curve states that contractors have learned how to repeat their mistakes.
- The successful contractor must stay vigilant to identify recurring mistakes and avoid them in the future.
- Unsuccessful (or ex-) contractors usually fail because of a pattern of behavior, not because of an isolated failure.

Figure 2.21

Management Characteristics of the Successful Contractor

Management of In-house Pre-Job Conferences

- Successful contractors should have "pre-mortems" (or pre-construction meetings) to *prevent* the bad job, rather than waiting to assess it in a "post-mortem."
- Before the job begins, the "old man" and the project team sit down and discuss the job, including its:
    - risks
    - challenges
    - staffing
    - priorities
- The "old man" shares his wisdom; the project team shares its view — their understanding and approach to the situation.
- The "Frisby Learning Curve" problem is discussed and a commitment made to avoid it.

Construction People Run Construction Companies

- Experienced construction people should run construction companies.
- When financial types or lawyers start running a construction company, it is usually a signal of weakness in the "old man."

Operational Management

- No company is so large that its key executive personnel should become removed from or unfamiliar with its operations.
- Executives should periodically "touch the flesh" of each operation...of estimating, purchasing, job management, cost reporting, etc.

Management of Expectations

- Let others know what you expect of them. "Others" can be your personnel, subcontractors, architects, and owners.
- Let others know what they can expect from you.
- Manage others so that they live up to your expectations.
- Manage yourself so that you can live up to others' expectations of you.

Management of Motion and Time

- Manage the progress of functions.
    - moving the Invitation for Bid through estimating to the successful low bid;
    - moving equipment from the description in the specs to effective installation and operation;
    - moving personnel successfully through the ranks and avoiding the "Peter Principle" (promoting people to their level of incompetence).
- Manage the movement of functions within just the right time frames.

Management of Finances Requires:

- a solid financial plan;
- a good knowledge of costs;
- accurate and honest cost reporting by the field;
- return on investment objectives;
- an honest relationship with a bank and surety;
- an excellent C.P.A.

Figure 2.21 (continued)

is impossible to be insured against incompetent or ill-prepared management. The only way for a contractor to protect the firm from this latter kind of damage is to develop and insist upon a proactive management style in its organization.

Summary

Neither successful nor unsuccessful companies just happen. Both are involved in a process of management. Management of the successful company identifies and manages risks, knows the market and market conditions, has a solid organization built around PPOIC (see Figure 2.16), and has a solid financial management plan. The unsuccessful company lets things happen rather than anticipating and controlling events.

The successful company *manages* the market, whereas the unsuccessful company *reacts* to the market. The successful company *manages risk*, whereas the unsuccessful company is generally a *gambler*. The successful contractor plans growth and grows moderately over the years, tying his organization together through PPOIC management, with input from, and expectations communicated to, all personnel. The unsuccessful contractor hires off the street and keeps mumbling about how bad communication is within the company. The successful contractor *plans* for organization, work acquisition, projects, finances, and personnel development—whereas the unsuccessful contractor just deals with these issues as they come along.

The successful contractor can *choose* between success and failure, whereas the unsuccessful contractor leaps without thinking to take advantage of apparent opportunities, without a strong organization behind him. This chapter is not written to provide the answers, but to cause the president and his staff to think about the questions and possible solutions *before* the questions are answered *for* them.

Chapter Three

Bidding and Estimating

Chapter Three

Bidding and Estimating

This chapter is concerned with matching the company's *estimate* of its ability to perform work with its *actual* ability to perform that work. The concept here is to develop, through the checklists, a discipline for:

- Determining whether to bid or not bid a project
- Identifying the risks in a construction project and dealing with those risks
- Continuing to monitor your company's performance by using the Bid/No Bid meetings as an opportunity for current and ongoing evaluation of how the company is faring
- Gathering the necessary information to carry out the goals listed above

This chapter deals not with how to estimate, but rather the information that must be considered and the risk assessment that must be performed as a part of the estimating process.

A well-known construction consultant once said: "I've never known a contractor to go out of business because of the job he did *not* get. However, I've known plenty of contractors who went under because of the jobs they did get." The decision to bid is one of the most important a contractor makes. It fulfills two functions:

- It provides the contractor an opportunity to match the capability of the company to the proposed project. This means considering the bid/no bid question: "Does our company have the capability to perform this job and complete it on time and with a profit?"
- It provides a first step in risk identification and management. A major responsibility of every construction company president is *risk management*. The bid/no bid decision is an opportunity to recognize and evaluate risks, and to ensure that all levels of the company are following suit.

Note: For more information on specific estimating functions, see Estimating for the General Contractor, Unit Price Estimating, or Means Square Foot Estimating, all published by R. S. Means. Means also publishes a variety of specialty "how-to" estimating references for various trades, including mechanical, electrical, landscape, steel, interior, and repair and remodeling, as well as a comprehensive handbook of charts and tables for use by estimators.

Identifying Risk

Risks fall into several categories as illustrated in Figure 3.1. There are many other types of risk that affect the construction industry. The point is that they must be *recognized*; the bid/no bid decision is a good place to begin that identification process.

Once the risk is identified, management must make a decision as to how to handle it. Some of the options include:

- *Avoiding the risk*, such as not bidding a project which is far larger and more complex than any others in which the contractor has experience.
- *Pricing the risk*, such as increasing the estimated labor cost on that remote project where skilled craftsmen are in short supply.
- *Managing the risk* by intense pre-planning and analysis so that additional cost can be avoided. Another option is subcontracting high risk work to a qualified, bonded subcontractor.

In deciding which of these three approaches to take, management should answer the questions set forth in the Bid/No Bid Decision Checklist and develop the information indicated (see Figure 3.5 later in this chapter).

Categories of Risks	
General Risk	**Examples**
Technical or Operational	• Dewatering • Tight tolerances • New techniques, such as slip forming
Contract	• Poor plans and specs • No damages clause • Exculpatory clauses
Time	• Constructive acceleration • Liquidated damages
Force Majeure	• Floods • Weather
Labor	• Lack of labor pool • Unqualified craftsmen • Strikes
Decision-Making	• Slow or unreasonable response by owner/consultant
Financial	• Slow pay • General economic conditions • Bankruptcy (Instability)

Figure 3.1

The estimate is an attempt to price the construction company's capability to perform this project profitably. This is the attitude the company must take—that the goal is to acquire work that can be performed profitably. Any time the attitude is: "I've got to have this job," trouble is inevitable.

The estimating phase is really the first pre-planning function, for no matter how large or small the project, the estimate must reflect how the job is to be built. Anyone can do a quantity survey and multiply the quantities times a historical labor factor. The genius behind successful estimating is to know and compare different methods of how the job should be built, the risks of each method, and what the company's capabilities are in this type of project.

It is generally believed that successful contractors do not bid everything that comes down the road. They bid selectively on projects of a type in which they are experienced, and they spend a bit more time on the pricing function—to ensure accuracy and to take into account a range of influencing factors. "Influencing factors" refers to the fact that each construction project is unique. Unlike an automobile factory which turns out thousands of repetitive copies of a given model car in a controlled environment, a given *construction project* is inevitably *different*. These "differences," which may involve location, supervisory team, how it is to be built, standards of acceptance, quality of drawings, etc., will influence the cost of construction. So, these factors should also influence the estimate of the cost of construction.

The Decision to Bid

The first and most important phase in the construction process is the *decision to bid*. Most bad jobs can be traced to a bad decision to take on a job that the contractor is simply not qualified for because of lack of personnel, experience, finances, or an impractical location, or the contractor just has too much going on at the time.

The president of the company should establish a "Bid/No Bid Decision" committee, usually comprised of him or herself as the chief estimator, the company's lead operational person, and the in-house accountant. A summary of the market plan should always be developed by and available to the committee to ensure that the criteria described in Chapter 2 (see Figure 2.5) are considered. Figure 3.2 shows the participants and the criteria involved in the Bid/No Bid decision, as well as the appropriate staff and corporate input. Figures 3.3 and 3.4a-c also show the chronology and mechanics of the decision-making process, using an example project and a hypothetical construction firm.

If this exercise is approached realistically, a risk can be analyzed and an informed decision made to handle that risk in one of the following ways:

- First, of course, it should be *avoided*; if this is not possible, then
- It can be *properly priced.*
- It can be *contracted out.*
- It can be reduced by application of *technical/operational/ management skill.*

In every case, you must be sure you are analyzing the *right risk*. In the example illustrated in Figure 3.3, perhaps it was not so much the job, but the contractor that presented the major risk. The "Checklists" section of this chapter contains a "Bid/No Bid Decision Checklist" that construction firms may use to ensure that all factors are considered in their own project analysis.

Estimating for the Acquisition of Profitable Work

As Franklin Roosevelt said: "In order for a person to be a *good* president, first he must *become* president." The same applies to the construction industry, where in order to make a profit on a construction project, first the company must get the job. Thus, the estimating phase is really the *acquisition* phase. The contractor's goal should be the *acquisition of profitable work*, not just being the low bidder.

The Estimating Function

After the decision to bid, the second most important phase is the estimating or pricing phase. Bidding and estimating are key to the health and success of a construction company. If the contractor performs the estimating or acquisition phase properly, it is because he is giving attention to the areas shown in Figure 3.5.

The estimator must ask these questions:

- Can *our company perform this work?*
- How much will it cost *our company* to do *this work?*

Participants and Input for the Bid/No Bid Decision		
Committee President Vice President of Operations Secretary-Treasurer Estimator		
Corporate Input	**Criteria**	**Staff Participants**
Market Plan	Job Size	Estimator/Project Manager
Budget/Goals	Type Job/Complexity	Estimator/Project Manager
	Location	Estimator/Project Manager
	Labor/Equipment Supply	Estimator/Bonding Agent
	Owner/Architect Reputation	Estimator/Project Manager
	Owner/Financial Integrity	Bonding Agent/Bank
	Percentage of Labor	Estimator/Accounting
	Competition	Bonding Agent
	Bid Trends	Estimator
	Profitability Trends	Accounting

Figure 3.2

An Example of the Bid/No Bid Decision Process

Corporate Input	Criteria	Staff Input
Market Plan shows that company's capability and goals are in the $500,000-$2,000,000 range.	Job Size	Estimator indicates that this is a $3,000,000 project.
Market Plan indicates that the contractor is interested in breaking into plant work.	Type Job/ Complexity	This is a sewer plant — remodel and extension. High percentage of sophisticated mechanical controls. Extensive mechanical excavation around existing plant.
Market Plan indicates that the contractor is seeking work within a hundred mile radius.	Location	In a major city fifty miles from contractor's principal place of business.
Market Plan aims at high rate of labor on this type of work to maximize profit.	Labor/ Equipment	Bonding Agent determines that other contractors in the area have found the labor market fairly tight. Control equipment, rebar and concrete are in short supply. Difficult to sub out mechanical excavation. Company's cash flow is tight.
Corporate Policy to be cautious in dealing with unreasonable owners or their representatives.	Owner/ Architect	One previous experience with Owner. Project Manager received letter of commendation. Architect has reputation for sloppy work — but has generally been reasonable to work with.
Corporate Policy to *not* bid financially irresponsible owners.	Owner's Financial Integrity	Owner is a major city — funds are available. Despite previous favorable experience, final payment is almost invariably delayed several months after final acceptance.
Market Plan is to look for better opportunities to increase profit — not just get another low return job.	Competition	Bonding Agent has learned that a total of only three general contractors are to be bidding—one the XYZ Corp. which is in financial difficulty and may be unable to bond. Looks like it could be a "fat" job for the successful contractor.
Corporate policy is to apply company resources to the projects that experience shows have the greatest likelihood of both acquiring and making a profit...don't bid where we haven't been successful.	Bid Trends	Estimator presents the following bid results by job size for the last quarter. These statistics are generally illustrative of the last several years.

Job Size ($)	Bid		Low Bidder	
	No.	%	No.	%
0 to 100,000	2	6%	2	18%
100,000 to 500,000	7	21%	2	27%
500,000 to 1 million	4	12%	2	18%
1 to 2 million	13	39%	4	36%
2 to 3 million	5	15%	0	0%
3 to 4 million	1	3%	0	0%
4 to 5 million	1	3%	0	0%
	33	100% (rounded)	10	100% (rounded)

Corporate Input	Criteria	Staff Input
Corporate Goals are for 3% return on sales against $10,000,000 billings.	Profitability Trends	Accounting indicates that on previous plant job, company grossed 3½% on sales, and that negotiated office buildings have brought in about 1½% gross on the average.

Other corporate tools would be available at the bid/no bid committee meeting. Figure 3.4a is a graphic presentation of sales goals versus actual sales. It represents that the company needs work. Figure 3.4b presents the status of gross profit earned to date against projections, also demonstrating that the company is well below its profit goals. Figure 3.4c shows that overhead is continuing at the same pace and is not being adequately absorbed by enough work in process. Now, with the data set forth above concerning the plant job, *how would you decide* (bid or no bid) if nothing else was before your organization to bid except two office buildings (about $1,000,000-$1,500,000) and one $700,000 multi-family project?

Figure 3.3

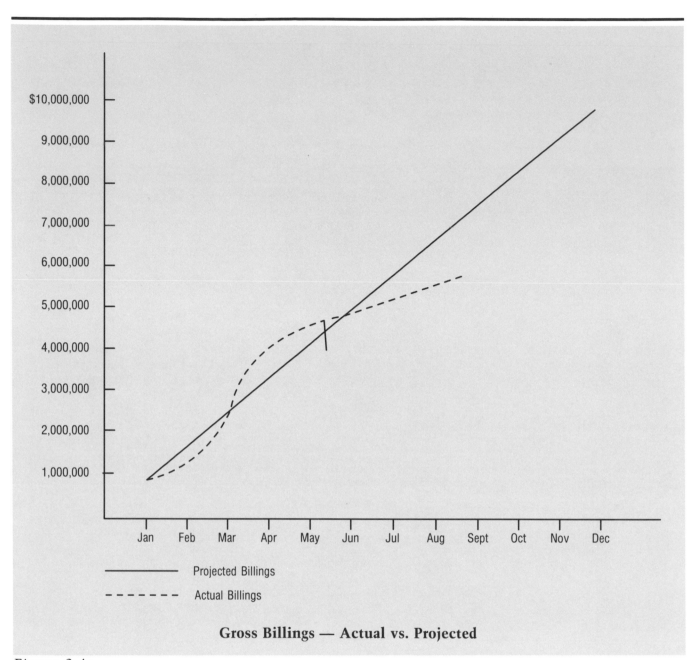

Gross Billings — Actual vs. Projected

Figure 3.4a

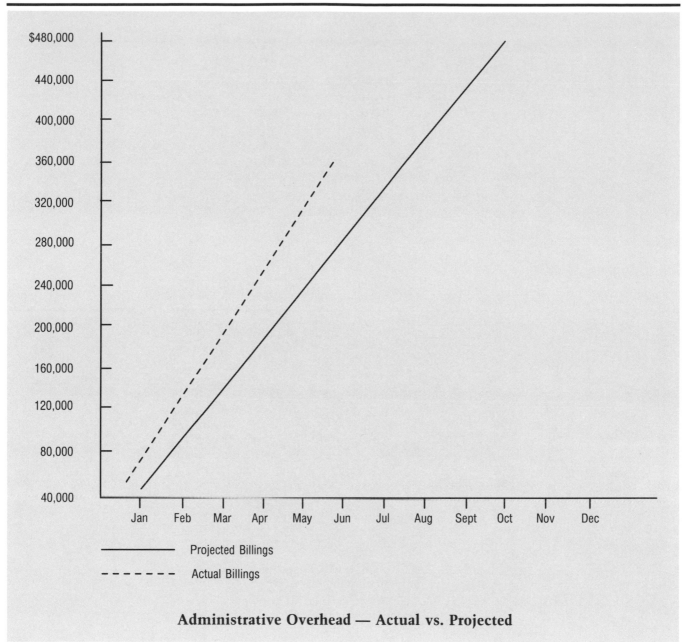

$480,000

440,000

400,000

360,000

320,000

280,000

240,000

200,000

160,000

120,000

80,000

40,000

Jan   Feb   Mar   Apr   May   Jun   Jul   Aug   Sept   Oct   Nov   Dec

——————   Projected Billings

- - - - - -   Actual Billings

Administrative Overhead — Actual vs. Projected

Figure 3.4b

- Are there other methods to complete this work faster or at less cost?
- Have we *identified* and *covered* the risks?
- How much *profit* should we make on this job?

Your Company, This Project

The words *your company* are emphasized because each company is unique, with different levels and types of expertise and experience than the competition. *This project* indicates a separate project, different from the last project. The estimator must know more than just the quantities and historical labor units; he must also know his company's strengths and weaknesses. He must

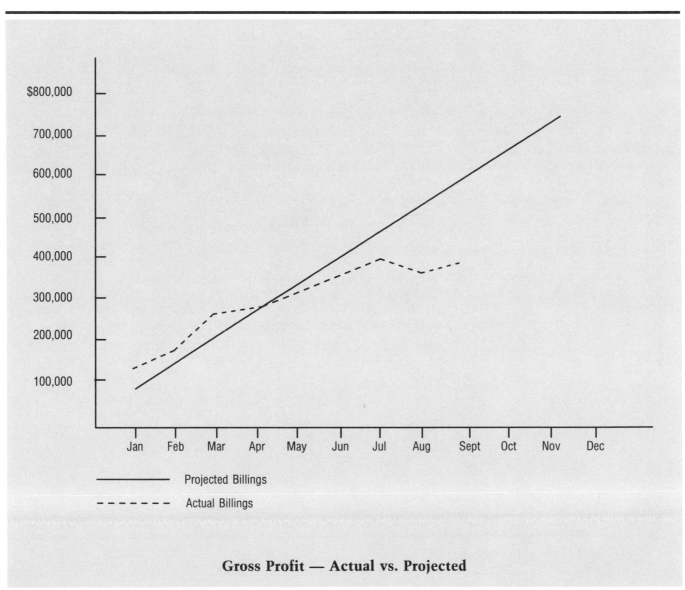

Gross Profit — Actual vs. Projected

Figure 3.4c

know and be able to price the *differences* in this project from other projects. Examples of possible differences include:

- The technical aspects
- Methodology
- Location of the project
- An uncooperative owner or owner's consultant
- Incomplete plans
- A financially weak owner
- Your own supervisory personnel
- Your own workload
- The labor supply
- The schedule
- The nature of the contract

Documentation

One of the major problems that arise in the estimating phase is a lack of documentation. It is important to note that proper documentation of the bid estimate becomes very important information to the field personnel who will be managing the project in general, to subcontractors and suppliers, and the pricing of changes and claims. The Estimating Checklist shown in the following "Checklists" section is a tool that can be used to ensure that all concerns set forth above have been properly addressed.

Bidding and Estimating Checklists

Bid/No Bid Decision Checklist

The Bid/No Bid Decision Checklist (Figure 3.6) is not a "one-shot" document. It is intended to be used to develop information about the potential job—from the time the plans and specifications come in—all the way until the bid recap.

Initially, the Bid/No Bid Decision Checklist is developed to provide management with enough information to determine whether (a) it should spend its resources estimating this job, or (b) it would be better off applying its resources to a job more compatible with the company's market plan. In that light, information will be developed to let management know (at the Bid/No Bid Decision Committee meeting) what this job is about, some of the risks that may be unusual, and whether this job is a type that the company should be pursuing.

After this initial information gathering, the checklist should be maintained on an ongoing basis. As the estimator gets deeper into the estimate and learns more about the job's risks and opportunities, the checklist should be updated accordingly. Certainly, at bid closing, the information set forth in the Bid/No Bid Decision Checklist should be made available to the construction firm's president, so he has a complete understanding of the risks, complexities, differences and opportunities in this potential project.

Pre-Bid Site Investigation Checklist

Over 20-25% of all changes and claims involve differing site conditions. It is not unusual for owners and architects to resist contractors' claims for changed conditions. This is because they

Estimating Functions and Requirements

Quantities

☐ Accuracy

☐ Completeness

Pricing

☐ Historic labor experience

☐ Schedule

☐ Winter work

☐ Production flow or crew sequence

☐ Field input

☐ Recognizing differences

☐ Identifying risk

☐ Available personnel

☐ Recognizing alternate construction methods

☐ Use of proper labor and other rates

Profit

☐ Risks

☐ Return on assets

☐ Economic conditions

☐ Duration of project

☐ Competition

Documentation in Bid File

☐ Pricing documents

☐ Schedules

☐ Quotes

☐ Pre-bid meetings, memos of interpretation

Job Site Overhead

☐ Experience of staff

☐ Project engineering needs

☐ Contract management needs

Figure 3.5

Estimating Functions and Requirements

Job Site Overhead (continued)

☐ Tools and moving equipment

    ____ lease vs. purchase

    ____ condition as a factor in productivity

☐ Quality control requirements

Home Office Overhead

☐ Certified payrolls required

☐ Other extra burden on home office

☐ Backlog that may affect overhead rate

Figure 3.5 (continued)

Bid/No Bid Decision Checklist

This original form will be started during the pre-estimate phase. Additional information will be entered on this checklist. When the information is entered, the responsible person will initial and date the entry. This original form should be reviewed during the estimating phase and bid closing, then filed and made part of the permanent record.

OUTLINE

A. General   E. Considerations
B. Evaluation Dates   F. Management/Estimating Concerns
C. Project Information   G. Bid Close Recap
D. Contract

A. GENERAL

   1. Project _____

   2. Owner _____

   3. Bid Date _____

B. EVALUATION DATES

   1. Pre-Estimate _____

   2. During Estimate _____

   3. Bid Closing _____

C. PROJECT INFORMATION

   1. Type Job (Housing, Commercial, P&I, Public Work, Other)

   2. Project Description: _____
_____

   3. Size (Approximate)

     $500,000 $1,000,000 $2,500,000 $5,000,000

     $10,000,000 $25,000,000 $50,000,000 $100,000,000

   4. Location: _____

   5. Customer (Name) _____

	Yes	No
a. Known by us	_____	_____
b. Prior projects profitable	_____	_____
c. Strong financial integrity	_____	_____
d. Reasonable		
e. Timely decisions	_____	_____

Figure 3.6

76

Bid/No Bid Decision Checklist

6. Architect/Engineer (Name) _____

	Yes	No
a. Known by us	_____	_____
b. Prior projects profitable	_____	_____
c. Strong financial integrity	_____	_____
d. Reasonable	_____	_____
e. Timely decisions	_____	_____
f. Typically provides adequate drawings & specifications	_____	_____

D. CONTRACT

1. Plans & Specifications

	Yes	No
a. Complete	_____	_____
b. Percentage complete _____		
c. Value engineering requested	_____	_____
d. Appear to be adequate	_____	_____
e. Unusual technical considerations	_____	_____
If so, what are they? _____		
f. Unusual aesthetic problems	_____	_____
If so, what are they? _____		
g. Prior project plans/spec problems?	_____	_____

2. Terms and Conditions

	Yes	No
a. Performance specifications (See checklist)	_____	_____
b. No Damages for Delay Clause	_____	_____
c. Fast track project	_____	_____
d. Special provisions _____		
_____		
_____		
_____		
e. Type of contract (Lump Sum, GMP, etc.) _____		

Figure 3.6 (continued)

Bid/No Bid Decision Checklist

        f. Savings split _____ _____ _____

        g. Bond requirements _____ _____ _____

        h. Current bond capacity _____ _____ _____

        i. Special insurance requirements _____ _____ _____

        j. Liquidated damages _____ _____ _____

        k. Government requirements (MBE, EEC, etc.) _____ _____ _____

    3. Financial

        a. Unusual financial arrangements _____

        b. Retainage _____

        c. Stored material _____

        d. Payment terms _____

        e. Lender _____

        f. Change order fees _____

		Yes	No
4. Schedule			
a. Liquidated damages reasonable		_____	_____
b. Actual damages		_____	_____
c. Reasonable schedule		_____	_____

        d. Contractual scheduling requirement _____

        e. Contract duration _____

E.   CONSIDERATIONS

		Yes	No
1. Personnel			
a. Estimating	Very overloaded now	_____	_____
b. Staffing personnel	Strong in-house capacity	_____	_____
c. Labor	Skilled workmen available	_____	_____
d. Subs	Qualified subs to bid	_____	_____

        If answer is no, which sub trades are weak?_____

        _____

        _____

        _____

        _____

Figure 3.6 (continued)

Bid/No Bid Decision Checklist

2. Competition
   a. Current number of bidders _____
   b. Anticipated number of bidders _____
   c. Probability of getting the job _____

		Yes	No
3.	Experience with this type of project		
	a. Typically makes profit	_____	_____
	b. Usually high change order rate	_____	_____
4.	Need for this job		
	a. Consistent with market plan/financial plan	_____	_____
	b. We need it so much that we'll risk a loss	_____	_____

ADVANTAGES   DISADVANTAGES

_____   _____
_____   _____
_____   _____
_____   _____
_____   _____
_____   _____
_____   _____
_____   _____
_____   _____
_____   _____
_____   _____
_____   _____
_____   _____
_____   _____
_____   _____
_____   _____
_____   _____
_____   _____

Figure 3.6 (continued)

Bid/No Bid Decision Checklist

A. MANAGEMENT/ESTIMATING CONCERNS

   1. Representations by owner or architect during negotiation

      _____

      _____

      _____

      _____

   2. "Hookers" in contract (i.e., clauses which are high risk, or which try to unreasonably transfer risk of owner's or architect's performance to contractor)

      _____

      _____

      _____

      _____

   3. Unusual technical provisions

      _____

      _____

      _____

      _____

   4. Labor considerations

      _____

      _____

      _____

      _____

   5. In-house staff capabilities

      _____

      _____

      _____

      _____

   6. Owner's involvement in estimate and with subs

      _____

      _____

Figure 3.6 (continued)

Bid/No Bid Decision Checklist

7. Location of site

   _____
   _____
   _____
   _____

8. Soils report and pre-bid site investigation

   _____
   _____
   _____
   _____

9. Financing

   _____
   _____
   _____
   _____

10. Schedule

    _____
    _____
    _____
    _____

11. Interface problems (joint venture, separate contractors, divisional interface, etc.)

    _____
    _____
    _____
    _____

12. Project team considerations

    _____
    _____
    _____
    _____

Figure 3.6 (continued)

Bid/No Bid Decision Checklist

13. Is the current status of estimating checklist acceptable?

Initials _____ Date _____

G. BID CLOSE RECAP	Amount	% of Total
a. Direct Labor	_____	_____
b. Direct Material	_____	_____
c. Subcontracts	_____	_____
d. Subtotal	_____	_____
e. General Conditions	_____	_____
f. Management Reserve (Contingency)	_____	_____
g. Fee	_____	_____
TOTAL	_____	_____

Figure 3.6 (continued)

believe contractors should learn all they can about the physical conditions of the site as a part of their duties under the Pre-bid Site Investigation clause.

The Pre-Bid Site Investigation Checklist (Figure 3.7) provides the contractor with a tool for gathering and evaluating all information required under the Pre-Bid Site Investigation clause. This checklist ensures not only a complete investigation, but also thorough documentation of the information gathered. A more detailed form, "Means Spec-Aid," appears in the Appendix. This form, organized according to the CSI (Construction Specifications Institute) MASTERFORMAT divisions, can be used as a checklist to ensure that no item is inadvertently excluded.

Estimating Checklist

Proper handling of the estimating phase is crucial to the health of a construction company. If the estimating is performed correctly, the contractor has a good shot at both getting the job and performing it profitably. If the job is not properly estimated, the contractor may lose the bid, or worse, win the bid and lose money on the project. In reviewing the bid documents, the estimator must consider the following:

- How are we going to build this job?
- What are the risks?
- How does this job differ from our usual work?
- What are our company's special advantages?
- Who do we have available to build it?
- Have I covered all the bases?

Many other questions must also be answered, and no checklist can be 100% complete for every company. The purpose of the Estimating Checklist in Figure 3.8 is to:

- Ensure that the Bid File is complete and adequately documented.
- Ensure that the estimating team has *thought* about the risks, methodologies, the company's advantages, and opportunities. Hopefully, they will also have discussed these issues so that the estimate is a "blueprint for victory," and not just another exercise in cranking out numbers.

Contract Documents Checklist

More and more, owners and their consultants are finding new ways to transfer risk to their construction contractors. Often, this "risk shift" takes place in the General or Special Conditions. In some cases, however, risk is shifted through the wording of the technical provisions. A risk shift is actually a financial shift, in that the owner is stating his/her unwillingness to bear this particular financial risk, which now belongs to the contractor. Sometimes these risk shifts are obvious and straightforward, such as very clear design and build (performance) specifications. However, they can also be subtle or ambiguous, such as in the case of some hybrid performance specifications or so-called No Damage for Delay clauses.

Pre-Bid Site Investigation Checklist

A. GENERAL

   1. Existing Site _____

      (brief description) _____

      a. Soil Condition _____

      b. Vegetation _____

      c. Access _____

      d. Existing structures _____

   2. Availability of Facilities

      a. Power _____

      b. Water _____

      c. Telephone _____

      d. Gas _____

      e. Adjacent vacant land _____

      f. Emergency services _____

      g. Roads (traffic, restrictions)

   3. Subsurface Conditions

      a. General description _____

      b. Rock _____

      c. Blasting required (yes) (no) _____

      d. Water _____

      e. Others _____

	Initials	Date
4. Soil Report Obtained A/E	_____	_____
5. Data		
a. Performed our own site investigation	_____	_____
b. Checked with local soils engineer	_____	_____
c. Taken pictures of site	_____	_____
d. Obtained weather data	_____	_____
e. Made inquiries of owner regarding any questions we may have	_____	_____

Comments: _____

_____

Figure 3.7

Pre-Bid Site Investigation Checklist

B. SURFACE CONDITIONS	Yes	No
1. Is site accessible?	_____	_____
2. Are all utilities available?	_____	_____
3. Are other contractors working in area?	_____	_____
4. Will access roads require maintenance?	_____	_____
5. In the immediate area, is there:		
a. A stream?	_____	_____
b. A dam?	_____	_____
c. A structure which will affect construction operations (such as a nearby building which could affect crane operations, etc.)?	_____	_____
6. Is borrow area easily accessible? How far?	_____	_____
7. Are there any special safety problems (highlines, etc.)?	_____	_____
8. Are elevations apparently as represented in the drawings?	_____	_____
9. Any activities or operations by owner that can affect construction work?	_____	_____
10. Is adequate space available for equipment, material storage, field office?	_____	_____
11. Is your observation of site consistent with conditions shown on plan and drawings?	_____	_____
12. Is site subject to flooding?	_____	_____
Comments: _____		

C. SUBSURFACE CONDITIONS		
1. Are there sufficient test holes shown in bid documents?	_____	_____
2. Is there sufficient data about subsurface conditions shown in bid documents?	_____	_____
3. Have we drilled our own test holes?	_____	_____
4. Have special measures been taken for:		
a. Dewatering?	_____	_____
b. Protecting against flooding during construction?	_____	_____
c. Blasting of rock, over-excavation, etc.?	_____	_____
d. Shoring of ditches?	_____	_____

Figure 3.7 (continued)

Pre-Bid Site Investigation Checklist

	Yes	No
5. Are underground utilities identified in bid documents?	_____	_____

6. Do the bid documents contain:

	Yes	No
a. A rock clause?	_____	_____
b. A clause compensating you for flooding or other acts of God?	_____	_____
c. Other protections against contingencies?	_____	_____

7. List contingencies or risks that must be considered in bid prices.

Comments: _____

_____

	Yes	No
8. Did your investigation disclose any inconsistencies with conditions shown in plans and drawings?	_____	_____

Comments: _____

_____

D. PERMITS, FEES AND TAXES

1. List permits and fees required and show cost (if applicable).

Item	Cost
a. Building Permit	_____
b. Electricity	_____
c. Water	_____
d. Gas	_____
e. Telephone	_____
f. Other	_____

2. Is the project tax exempt? _____

3. If not tax exempt, show applicable taxes.

   a. State Sales (Use) Tax _____ %

   b. Local Sales (Use) Tax _____ %

   c. Other _____ %

_____

	Yes	No
4. Are there other special fees or licenses that will be required? (List)	_____	_____

Comments: _____

_____

Figure 3.7 (continued)

Pre-Bid Site Investigation Checklist

	Yes	No
E. LABOR		
1. Is labor readily available?	_____	_____
2. Will overtime/shift work be required?	_____	_____
3. Are there other major projects coming up in the area which could affect labor market?	_____	_____
4. Are labor rates due to increase (new union agreements, etc.)?	_____	_____
5. Are strikes possible due to expired labor contracts?	_____	_____
6. Is project being performed in a geographical area or under conditions which could cause inefficiencies?	_____	_____
7. Will travel time be required?	_____	_____
8. Are there any special potential jurisdictional problems?	_____	_____

Comments: _____

_____

F. WEATHER

	Yes	No
1. How much down time can you expect based on weather (e.g., 10 year climatological report from weather bureau)?	_____	_____
2. Rainfall	_____	_____
3. Must special measures be taken to protect against:		
a. High Winds	_____	_____
b. Flooding	_____	_____
c. Rain	_____	_____

Comments: _____

_____

G. EQUIPMENT

1. List special equipment that may be required for the project.

Comments: _____

_____

2. Is nature of site remote, difficult terrain (such that maintenance, depreciation factors must be adjusted)?

Comments: _____

_____

3. List any special problems with transporting equipment to job site.

Comments: _____

_____

Figure 3.7 (continued)

Pre-Bid Site Investigation Checklist

	Yes	No
H. TRANSPORTATION		

1. Does schedule dictate special transportation (air freight, etc.) _____ _____

2. Any pending transportation strikes? _____ _____

3. List any special transportation problems.

    Comments: _____

    _____

    _____

4. Provide transportation for workers? _____ _____

    Comments: _____

    _____

I. PRICES

1. Any price hikes (steel, plywood, etc.) pending? _____ _____

2. Any pending industry strikes which could escalate prices or cause material/equipment shortages and delays? _____ _____

    Comments: _____

    _____

J. MEETINGS

1. Summarize meetings, telecons, discussions with Owner or his representative concerning conditions.

DATE	CONTRACTOR REP.	OWNER REP.	SUMMARY

Figure 3.7 (continued)

Estimating Checklist

A. BID DOCUMENTS	Initials	Date
1. Contract	_____	_____
2. Drawings	_____	_____
3. Specifications	_____	_____
4. Addenda		
From: _____ To: _____	_____	_____
5. Referenced Soil Report	_____	_____

B. BID FILE INCLUDES:		
1. All value engineering proposals (logged into Changes Log)	_____	_____
2. All telephone memos	_____	_____
3. All pre-approvals on materials (logged into Submittal Log)	_____	_____
4. All schedules	_____	_____
5. All take-off sheets	_____	_____
6. All sub quotes (logged into Subcontractor Log when appropriate)	_____	_____
7. All pricing sheets	_____	_____
8. Memos of all discussions with owner/representative		
9. Memos of all "interpretations" made by us	_____	_____
10. Requests for Clarification (logged into RFI Log)	_____	_____

C. ESTIMATE HAS BEEN:		
1. Checked for quantity errors	_____	_____
2. Checked for math & extension errors	_____	_____
3. Reviewed for pricing by:		
a. Field personnel	_____	_____
b. Division personnel	_____	_____
c. Purchasing	_____	_____
4. Checked for responsiveness	_____	_____

D. ESTIMATE CONSIDERATIONS		
1. A pre-bid site investigation was made in accordance with our procedure (See Pre-Bid Site Investigation Checklist, Figure 3.7)	_____	_____

Figure 3.8

Estimating Checklist

	Initials	Date
a. Reference reports were reviewed	_____	_____

b. The following are the potential site risks which we have considered:

1. Risk: _____

   Managed by: _____

2. Risk: _____

   Managed by: _____

3. Risk: _____

   Managed by: _____

4. Risk: _____

   Managed by: _____

	Yes	No
2. The Contract Document, or "Gotcha," Checklist (Figure 3.9) has been reviewed and taken into consideration. "Gotcha's" are unusual or high risk clauses contained in the General or Special Conditions or the Technical Specifications. A "Gotcha" Checklist follows the Pre-Bid Site Investigation Checklist.	_____	_____
a. Legal has reviewed the items identified above	_____	_____

3. Value engineering proposals have been:

	Yes	No
a. Clearly established as to scope of work	_____	_____
b. Coordinated with other sections of the contract	_____	_____
A list of value engineering proposals is attached.	_____	_____

4. The following are *differences* from the usual project and have been taken into account in the estimate:

a. Type of contract _____

_____

b. Geography _____

_____

c. Structural _____

_____

d. Special construction methods _____

_____

Figure 3.8 (continued)

Estimating Checklist

e. Methodology _____
_____

f. Quality or testing _____
_____

g. Status of drawings _____
_____

h. Tightness of schedule _____
_____

i. Fast track _____
_____

5. The following categories of labor have been priced *below* or *above* the usual unit cost:

Work Item	% +/- per Unit
_____	_____
_____	_____
_____	_____
_____	_____
_____	_____

6. Contingencies have been included for the following work items:

Work Item	Contingency
_____	_____
_____	_____
_____	_____
_____	_____
_____	_____

7. Considerations have been given to the following schedule items:

	Yes	No
a. Overtime has been included	_____	_____
b. Weather	_____	_____
c. Time-sensitive cost	_____	_____
d. Labor escalation	_____	_____
e. Contingency for liquidated/actual damages has been included	_____	_____

Figure 3.8 (continued)

Estimating Checklist

E. PROJECT STAFFING

	NAME	DURATION
1. Project Manager	_____	_____
2. Superintendent	_____	_____
	_____	_____
3. Field Engineer	_____	_____
4. Project Engineer	_____	_____
	_____	_____
	_____	_____
	_____	_____
5. Cost Engineer	_____	_____

F. POST BID DISCUSSION

1. Good points about this bid effort:

_____

_____

_____

2. Items needing improvement:

_____

_____

_____

This checklist has been prepared:

By: _____ As of: _____

Revised By: _____ On: _____

Revised By: _____ On: _____

Figure 3.8 (continued)

The Contract Documents Checklist (Figure 3.9) provides the framework for a disciplined review of the contract documents for both risk shifts and other unusual risks. With an organized method for recognizing and recording risks, it then becomes easier to decide how to handle them—whether to price the risks or manage them so as to avoid extra costs (always keeping one's eyes open, of course).

Summary

The plans and specifications that the estimator is pricing will become a major component of the contract if the bid is successful. The contract establishes the limits of the scope of work, as well as the risks the contractor is to undertake. The contract also establishes the risks that do not belong to the contractor—that is, the owner's risks. The estimator must be proficient and accurate at taking off quantities and estimating how many labor and/or equipment hours it will take *this company* to install those quantities. In addition, he or she must also, along with the other parties responsible for complying with the terms of the contract, be a skilled contract interpreter. Contract interpretation is the subject of the next chapter, and the next set of checklists.

Contract Document Checklist

A. THE SEARCH FOR "GOTCHAS" ("Gotchas" are unusual or high risk clauses contained in the General or Special Conditions, or the Technical Provisions of the bid documents.)

Clause	What To Look For

1. Differing Site Conditions
- ☐ Notice requirements (such as "immediately upon discovery of the changed condition")
- ☐ Unusual disclaimers
    - ___ In clause
    - ___ In soils report
- ☐ Reference to other soils reports (e.g., in A/E's office)
- ☐ By submitting bid, contractor warrants site conditions
- ☐ Absence of differing site conditions clause
- ☐ Inadequate soils report (e.g., "5 borings on 5 mile line.")
- ☐ No soils report
- ☐ Performance requirements (e.g., contractor bears complete risk for dewatering, irrespective of actual conditions.)
- ☐ Elevations determined by aerial photography

2. Delay Clause
- ☐ No damages for delay
- ☐ Time extension only remedy for delays
- ☐ Notice requirements
- ☐ Specific delays listed for which no costs are to be paid for time extensions:
    - ___ Change orders
    - ___ Owner-furnished equipment (OFE)
    - ___ Drawings released
    - ___ Building permits
    - ___ Other contractors on site
    - ___ Force Majeure

3. Change Orders
- ☐ No unilateral right to issue change
- ☐ Forward pricing of changes required
- ☐ As to Impact:
    - ___ Method of pricing changes is set forth

Figure 3.9

Contract Document Checklist

Clause	What To Look For
	_____ Is there a set overhead and fee rate?
	_____ Limitations on what can be priced (e.g., premium on overtime only)
	☐ Authority to issue change limited (as in some public works contracts)
	☐ Notice requirements strictly enforced
4. Project Completion	☐ Are the Substantial Completion requirements identified?
	☐ Partial owner turnover requirement
	☐ Defective work or materials repair
5. Payments	☐ Liens by general waived
	☐ Liens by subcontractors waived
	☐ Unusually lengthy time for payments
	☐ Any subordination
	☐ Final completion/release of retainages
	☐ Payment procedure
	☐ Stored material provision
6. Financing	☐ Review finance agreements
	☐ Notice requirements to lender regarding changes
	☐ Consistency with contract documents
	☐ Limitations on cost growth
	☐ Construction funding tied to sales, etc.
7. Liquidated Damages	☐ Amount unusual
	☐ Actual damages only
8. Schedule Requirements	☐ Fast track
	☐ Method requirement (e.g., CPM)
	☐ Unreasonable duration

Figure 3.9 (continued)

Contract Document Checklist

Clause	What To Look For
	☐ Update Requirements
	☐ Coordination of separate contractor
9. Claims	☐ Arbitration Clause
	☐ Authority of A/E in resolution
	☐ Government immunity

B. TECHNICAL PROVISIONS

Nature Of Risk	What To Look For
1. Transferring Risk	☐ All performance specifications
	☐ All specifications using words like "design," "engineer," "performance," to describe any part of the contractor's responsibility
	☐ All clauses placing "coordination of design or equipment interface"
	☐ All clauses attempting to shift liability for design error to contractor
2. Unusual Tolerance	Such As:
	☐ 1/8" in 10' on concrete
	☐ Extremely close rebar spacing
3. Non-Quantitative Requirements	Such As:
	☐ "Perfect weld"
	☐ "Crack-free concrete"
	☐ "Non-porous concrete"
	☐ "True and plumb"
	☐ "Even"
	☐ "Architect sole judge of contract compliance"
4. New Products, Testing Procedures	☐ Type K concrete
	☐ Hercules

Figure 3.9 (continued)

Contract Document Checklist

Nature Of Risk	What To Look For

	☐ Splice test with plumb bob
	☐ F-numbers for flat work
5. Ambiguities, Conflicts Between Contract Documents & Estimate	
	☐ Document contracts with owner
	☐ Include written memo as to how you bid it
6. Warranty Requirements	☐ Unusual warranty duration
	☐ Unusual scope of warranty
	☐ Start of warranty period
7. Subcontractors/ Suppliers	☐ Owner approval or selection
	☐ Ambiguity in scope of work as defined by specification
	☐ Experience requirements
	☐ Design or performance requirements
	☐ The decision not to bond a subcontractor
	☐ Contract scope of work consistency with local craft jurisdiction

C. REPRESENTATIONS DURING ESTIMATING PHASE

1. Made to Owner	☐ Schedule
	☐ Quality of construction
	☐ Scope of work included in estimate
	☐ Being able to handle incomplete drawings
	☐ Being able to handle numerous changes to the work
	☐ Personnel to be used
	☐ Use of subcontractors
	☐ Reports to be submitted

Figure 3.9 (continued)

Contract Document Checklist

C. REPRESENTATIONS DURING ESTIMATING PHASE

Nature Of Risk	What To Look For
2. Made by Owner	☐ Coordination of separate contracts
	☐ Completeness of design
	☐ Capacity to fund project
	☐ Quality of work expected
	☐ Authority of representative
	☐ Move-in schedules
	☐ Start date

Figure 3.9 (continued)

Chapter Four

Contract Interpretation

Chapter Four

Contract Interpretation

This chapter addresses the construction firm's interpretation of the contract, an essential element in any strategy for success. The checklists in this chapter provide guidelines for contract interpretation, a means of identifying prescription versus performance specifications, and a method for determining where No Damages for Delay clauses apply. Also covered are Pre-Bid Site Investigation clauses, Differing Site Conditions clauses, and scope of work interpretation.

Contract interpretation is a technical and risky business. The first step in contract interpretation is to know what constitutes the contract. In a bid situation where the owner advertises for contractors' bids, the contract is a composite of *express* rights and duties, and *implied* rights and duties. In fact, in order to gain an understanding of the overall contract, one would have to analyze the documents, legally enforced standards, and implied obligations, as shown in Figure 4.1. Remember, each contracted situation is different, and Figure 4.1 is only an illustration.

The Contract Base Line, or "Fence"

The contract between the contractor and the owner acts as a "fence," or boundary, which distinguishes between the *contractor's duties* (and what it must provide through the contract sum) and the *owner's duties* (and what it must provide to the project). The fence is also called the *contract base line.*

The fence is initially established during the estimating phase of the contract. It is therefore essential that the estimating staff be aware of the principles of the fence and thoroughly document the risks that must be priced as included in, and excluded from, the fence.

The project staff responsible for building the job must have the same understanding of the fence, or contract base line, as the estimating staff that priced the job. The reason this is so important is that a major function of the project manager's staff (by definition all personnel responsible for constructing the project) is to protect that fence. It is the duty of the project staff to provide the owner everything inside the fence that defines the contractor's responsibilities; but it is their equally important duty to guard against the owner, and the owner's representative or other party from encroaching upon that fence.

To ensure that the fence priced by the estimators is the same as that protected by the project staff, the project staff must be adequately briefed after the contractor has successfully bid a job. (A helpful tool for transferring this information from estimator to project staff is a completed Spec-Aid form, as shown in the appendix.) The profit center manager is responsible on every job to ensure constant monitoring of the integrity of the contract base line. The fence generally functions in arrangements similar to those shown in the figures that follow. However, the fence—and the contractor's role—may vary (in subtle or dramatic ways) from project to project. The estimating staff and the project manager

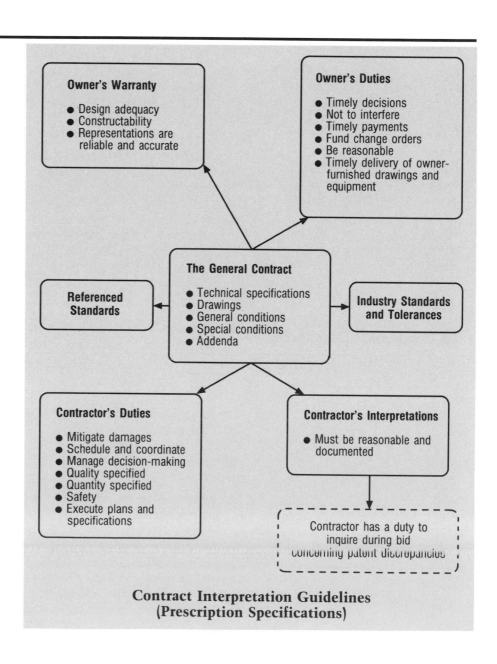

**Contract Interpretation Guidelines
(Prescription Specifications)**

Figure 4.1

must be constantly attentive to these differences, and aware that both the pricing and administration process captures and manages the uniqueness of each project.

Prescription and Performance Specifications

The Spearin Doctrine

In this fence (see Figure 4.2), the owner has prepared detailed plans and specifications, known as *prescription specifications*. Under the Spearin Doctrine, a federal, universally accepted case, the owner is responsible for warranting the adequacy of the design in prescription specifications. If the plans are defective, the owner bears the risk of that loss. According to this doctrine, the owner also has a duty to perform decision-making duties on a timely basis.

In prescription (or design) specifications, the owner and/or architect specifies *what* is to be done and *how* to do it. In performance specifications (see Figure 4.3), the owner only names the function he wants performed and leaves the *how* to do it to the contractor. In this case, the owner does *not* warrant the adequacy of the specifications.

Performance Specifications

One vehicle that owners use in an attempt to transfer the risk of inadequate plans and specifications is performance specifications (as shown in Figure 4.3). In a *true* performance specification, the risk for the work (or *how* to do it) is transferred to the contractor; therefore, the contractor's risk (or scope of work) is expanded—his fence risk is enlarged as shown.

Performance Versus Prescription Specifications Checklist

A simple description of performance and prescription specifications is given in Figure 4.4. However, in cases where it is difficult to determine whether the specifications are performance or prescription, one can use the Performance versus Prescription Checklist in Figure 4.4 as a guideline.

The characterization of a specification as a *performance specification* does not automatically cast the performance requirement upon the contractor. Characterizing a specification as prescription or performance depends upon an analysis of the following factors.

1. Does the contract require the contractor to warrant design (performance) or construction?
2. Does the contractor have freedom of design (constrained only by achieving the end result in accordance with "sound engineering practice")?
3. Does the contract specifically impose engineering and design duties upon the contractor?
4. What were the engineering and design duties of the architect/engineer?
5. How long was the project in design?
6. How long did the contractor have to bid?
7. Are fabricated articles described by manufacturer (source)? What control does the owner contractually reserve in substituting named sources, or in making changes to the configuration of the named source's product?

8. Is shop drawing approval reserved for the owner (or its representative)?
9. If shop drawing approval is reserved, at what level of detail is that approval exercised?
10. Was the contractor required to furnish Errors and Omissions insurance?
11. Are any performance requirements in the contract documents contained in:
    a. Catch-all language, such as "complete operable systems?"
    b. Disclaimers and exculpatory provisions? Or are they set forth clearly and explicitly as positive obligations of the contractor?

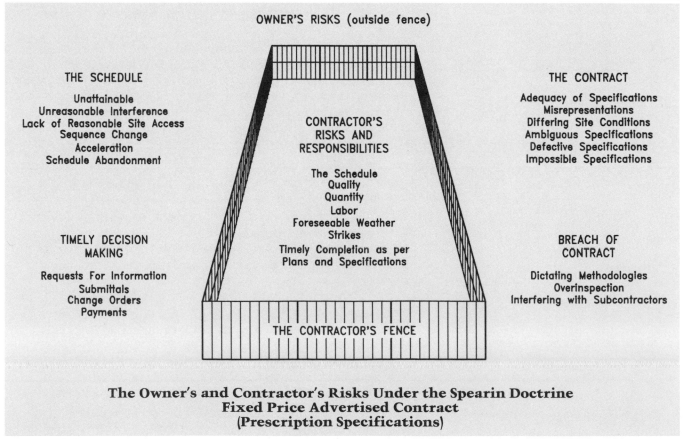

The Owner's and Contractor's Risks Under the Spearin Doctrine
Fixed Price Advertised Contract
(Prescription Specifications)

Figure 4.2

12. Under the contract documents, what specific design packages is the contractor required to submit?

If it appears from the foregoing analysis that:

- the architect/engineer contracted for the complete design of the project, and
- the architect/engineer selected the manufactured products to be used (purchase description and named source), and
- the contract documents do not require a performance or design warranty of the contractor, and
- the contractor had limited time to bid the project, and

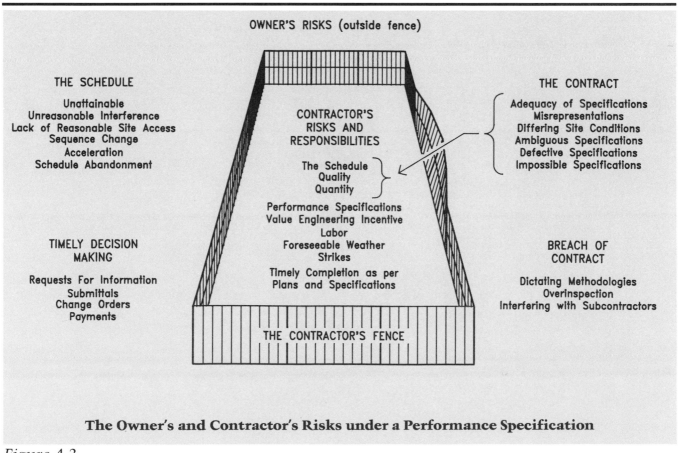

The Owner's and Contractor's Risks under a Performance Specification

Figure 4.3

- the contractor may not make changes to the approved sources or to the configuration thereof without owner approval, and
- the contract requires the contractor to submit shop drawings for approval, and shop drawing approval by the A/E is at a very low level of detail, and the contractor is unable to begin fabrication/manufacture without the A/E's approval, and
- all changes must be approved by the A/E and
- the A/E remains in a position of detailed control, inspection, and surveillance throughout the project . . .

. . . then it would be logical to conclude that the contractor does not have performance or design responsibilities imposed upon him/her contractually. The point is, of course, that determining whether specifications are prescription or performance requires a study of all the circumstances. In the event of a doubt, the contractor's interpretation (if reasonable) would prevail over the owner's.

The No Damages for Delay Clause

This clause is used by owners to shift the risk of loss to contractors for delays resulting from the owner's failure to make timely decisions, or from other delays to the contractor. Figure 4.5 shows how risk is "bounced" back and forth under a classic No Damages for Delay clause. Although the courts have held these claims to be legally enforceable, they are interpreted very strictly (i.e., their application may have narrow limits, and there are many exceptions). As a result, such clauses may not be as much of a shield to the owner as he may think.

For example, a No Damages For Delay clause may apply to the issuance of change orders (which puts the risk of a time extension

---

Characteristics of Performance vs. Prescription Specifications

In a Performance Specification...
- Contractor accepts responsibility for design and engineering.
- Contractor guarantees achievement of stated performance requirements.
- Owner reserves right of final inspection and approval.

A Prescription Specification...
- Provides precise measurements.
- Sets tolerance limits.
- Lists material to be used.
- States quality control requirements.
- Is similar to a Purchase Specification, which designates:
  — manufacturer and model number,
  — part number,
  — or specific product, sometimes modified by the phrase "or equal."

Figure 4.4

due to change orders inside the contractor's fence), but may then be deemed invalid if the delay is unreasonable or unforeseeable. At this point, the risk is bounced back to the owner.

The following is an example of a No Damages for Delay clause. The specific wording varies from contract to contract, but this example is typical.

> *"No charges or claim for damages shall be made by the contractor for any ordinary delays or hindrances, from any cause whatsoever, during the progress of any portion of the work embraced in this contract. Such delays or hindrances shall be compensated for by an extension of time as above provided."*

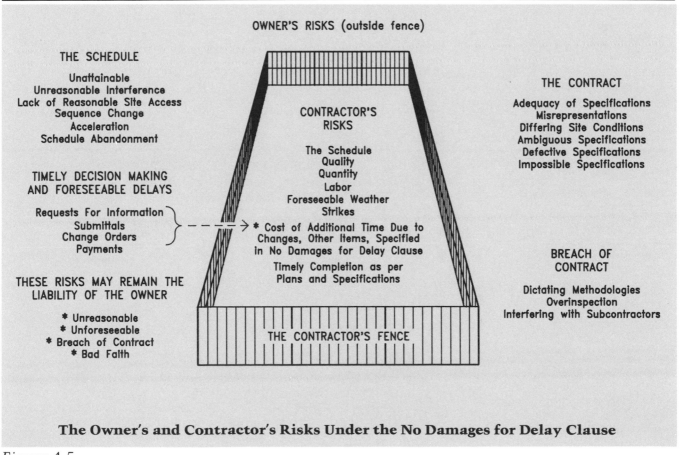

The Owner's and Contractor's Risks Under the No Damages for Delay Clause

Figure 4.5

Figures 4.6 and 4.7 provide guidelines for determining when the No Damages for Delay clause is enforceable, and when it is not. Figure 4.8 shows how a contractor should handle delay damages situations. Contractors should note that the No Damages for Delay clause may involve great risk, and should seek professional counsel.

The Differing Site Condition Clause

The Pre-Bid Site Investigation and Differing Site Condition clauses can be confusing for the following reasons. On the one hand, the Pre-Bid Site Investigation clause seems to impose on the contractor the complete risk for surface and subsurface conditions. On the other hand, the Differing Site Condition clause (in most contracts) protects the contractor from subsurface conditions that differ from those indicated in the contract or normally found in that area.

Figure 4.9 shows how the disclaimer or exculpatory clause in the pre-bid site investigation clause tries to shift the risk of loss for a differing site condition. In many cases, the Spearin doctrine and Differing Site Condition clause push the risk back to the owner's arena.

The Pre-Bid Site Investigation Checklist recommended for contractors is shown in Figure 3.7, Chapter 3.

Negotiated Contracts

The aforementioned "fences" were based on contractual arrangements where the project was advertised, and the contractor had no authority to vary the terms and conditions of the contract documents. Under the *negotiated* contract, the parties *may*

Determining When the No Damages for Delay Clause Is Enforceable

The "No Damages" clause is enforceable and applies where...

- Delays are foreseeable, such as:
  — Some RFI's
  — Some changes
  — Some delays in acting on submittals
  — Some decision-making delays
  — Some delays in site access

- Remember, though:
  — It is strictly construed because it is a harsh clause.
  — It applies only to delays.

- How to handle?
  — Either by pricing in the original estimate a labor and time contingency, *or*
  — pricing the delay cost as an impact on unchanged work
  — Requires more intensive project management changes.

Figure 4.6

Determining When the No Damages for Delay Clause Is Not Enforceable

The "No Damages" clause does NOT apply where ...

- There is an interference with the contractor's work, such as:
  - a sequence change
  - demobilization/remobilization
  - stop and go

- The owner fails to meet his/her contract duties, such as:
  - providing drawings on time (unless specifically covered in the "No Damages" clause)
  - providing owner-furnished equipment on time
  - providing site access on time

- Unreasonable delays are involved.
- Unforeseeable delays are involved.
- There is schedule abandonment.
- Acceleration is an issue.
- Constructive acceleration is an issue.

Figure 4.7

How to Handle Delay Damages Checklist

What to do:

☐ In original schedule:

____ Show "Production Flow"

____ Claim float as contractor's

____ Show change order cut-off dates (i.e., last point at which a change can be issued on each system without impacting schedule).

☐ If changes start flowing:

____ Try to convince owner to reduce the number of changes or to stop issuing them altogether in order to mitigate your damages, as he has a duty to do.

____ DOCUMENT ALL COMMUNICATIONS THOROUGHLY!

____ SHOW EFFECT ON CPM!

____ Show acceleration or sequence change.

Figure 4.8

Owner and Contractor Risks and Responsibilities Under the Pre-Bid Site Investigation Clause and Differing Site Condition Clause

Pre-Bid Site Investigation Clause:

☐ Imposes the following responsibilities on the contractor:

    ____ Visit site

    ____ See what should be seen

    ____ Review specs/drawings

    ____ Know labor market, non-physical conditions

    ____ Research weather conditions

☐ Does not require the contractor to:

    ____ Assume misrepresentation

    ____ Have or obtain knowledge of expert

    ____ Perform full soils investigation

Differing Site Condition Clause:

Type I

Owner Risks:

☐ Misrepresentation

☐ Spearin application

☐ Concealment

☐ Failure to issue direction

Contractor Risks:

☐ What he knew or should have known (standard industry applications, experience)

☐ Illogical, incomplete inferences

☐ Written notice

☐ Proof

☐ Failure to mitigate damages

Type II

Owner Risks:

☐ Unusual

☐ Unforeseeable

Contractor Risks:

☐ Previous experience in area

☐ Notice

☐ Proof

Figure 4.9

completely alter these rules and agree to place the financial responsibility for a given risk with any party on whom they mutually agree. Figure 4.10 shows that in situations where the contractor has some input regarding the contract terms and conditions or the plans and specifications, the owner's risk may diminish. The estimator must therefore be very careful of the negotiated contract and the risks that the company may inherit in the negotiation process.

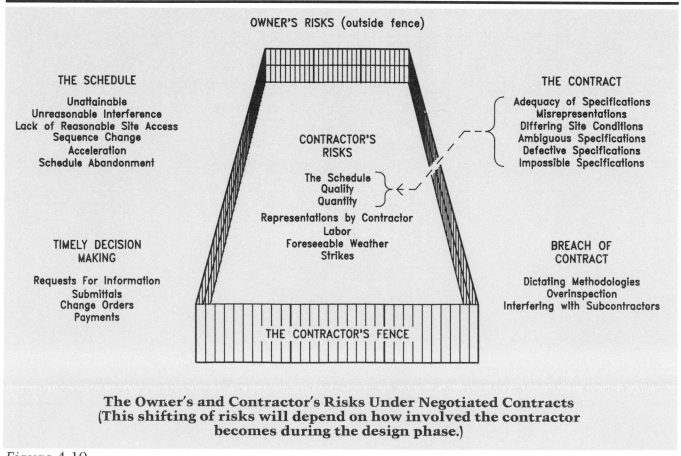

The Owner's and Contractor's Risks Under Negotiated Contracts (This shifting of risks will depend on how involved the contractor becomes during the design phase.)

Figure 4.10

Lack of Notice and/or Documentation

As Figure 4.11 shows, the contractor may waive some or all of its rights by failing to provide written notice of its claims or extras as presented by the contract, or by failing to maintain proper documentation. The estimator should be aware of these kinds of requirements and recognize that an adequate administrative staff is necessary to perform the absolutely essential paperwork tasks.

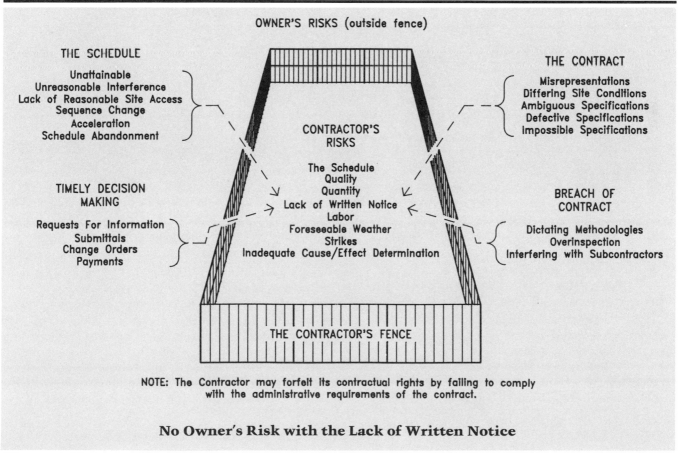

OWNER'S RISKS (outside fence)

THE SCHEDULE
Unattainable
Unreasonable Interference
Lack of Reasonable Site Access
Sequence Change
Acceleration
Schedule Abandonment

TIMELY DECISION MAKING
Requests For Information
Submittals
Change Orders
Payments

CONTRACTOR'S RISKS
The Schedule
Quality
Quantity
Lack of Written Notice
Labor
Foreseeable Weather
Strikes
Inadequate Cause/Effect Determination

THE CONTRACT
Misrepresentations
Differing Site Conditions
Ambiguous Specifications
Defective Specifications
Impossible Specifications

BREACH OF CONTRACT
Dictating Methodologies
Overinspection
Interfering with Subcontractors

THE CONTRACTOR'S FENCE

NOTE: The Contractor may forfeit its contractual rights by failing to comply with the administrative requirements of the contract.

No Owner's Risk with the Lack of Written Notice

Figure 4.11

Scope of Work Interpretation

Defining what is and what is not included in the Scope of Work and General Conditions starts in the estimating phase and continues throughout the job. A contractor must bid only that which is contained in the Scope of Work. The contractor cannot produce a competitive bid if he assumes that more work is expected than the scope actually requires. By the same token, the contractor subjects his company to cost overruns if he assumes that less work is necessary than what the scope actually covers.

The contractor's interpretation of the plans and specifications (including such phrases as "reasonably inferred") is vital to its profit and to its competitive position in the marketplace (see Figure 4.12). Therefore, it is important for the contractor to pay attention to the following rules of contract interpretation.

Rules of Contract Interpretation

Rule 1: Treat the Contract Documents as a Whole

A general contractor is responsible for all work described in a contract, wherever it occurs. This means that if electrical details are contained on the mechanical drawings, the general contractor is responsible for seeing that those details are met, even though it may have subcontracted the electrical as Section 13 (Electrical Specs) and the "E" drawings only. The issue then becomes: Who

---

How to Interpret Vague Terminology

"Reasonably Inferred"

Often contracts state that the contractor will provide the materials and do the work which is "reasonably inferred." What does this mean?

- This is not a "catch-all" for architect/engineer errors and omissions. It does not relieve the owner of the warranty of adequacy of design.

- It probably means that obvious details (like the manifest intent clause) are the contractor's risk (responsibility). For example, it might very well mean that if conduit is shown going from the transformer to the motor control center, the wire is to be inside the motor control center, or that cable goes on the cable tray. But if the motor control center is undersized, that is still the owner's responsibility. If fixtures are not shown in the ballroom, they are probably still the owner's responsibility, because it is an owner's choice item.

- In other words:
  — Is the item (or quality) an owner's choice? If yes, then it is probably not covered under the "reasonably inferred" clause.
  — If it is a mandatory construction installation item that any contractor should have picked up on, then it may be covered in "reasonably inferred."

- Another test:
  — If a bid (federal, state, municipal) job, the clause will be strictly construed against the owner.
  — If a negotiated job (especially CPGM), the clause will be more liberally construed against the contractor.

- Another test:
  — If you can show from your estimate that you bid an item, your interpretation will probably be accepted.

- Get professional advice; these are tough questions.

Figure 4.12

is accountable for the electrical detail on the mechanical drawings? Since it is listed *somewhere* on the drawings, the general contractor likely has no claim against the owner and may try to pass the responsibility on to one or more subcontractors. The general contractor can avoid this problem by:

- asking each subcontractor to bid on all the work in its craft (i.e., the mechanical subcontractor is to bid on all mechanical work, on whichever drawings are made available).
- ensuring that the subcontract agreement assigns the total work to that subcontractor.
- ensuring that the subcontract agreement clearly states the work that is included.
- giving close attention to the work where the mechanical and control systems meet or overlap. These problems are the most frequent and repetitive.

Rule 2: Make Reasonable Interpretations

Read the contract *reasonably* and ensure that the other parties do, too (see Figure 4.13). Keep in mind that the contractor should read this information with the perspective of a "reasonably prudent contractor," and *is not* bound by standards that require the knowledge of a consulting engineer. The contractor *is* bound by the knowledge and experience that it (and similar contractors) have had or should have had. This implies that the contractor's view of electrical drawings on an advertised federal project should be that the drawings are complete and adequate. For a power plant, however, it may be reasonable to assume that the drawings are incomplete and that some delays may occur in completing the drawings—and the contractor's bid should reflect it.

Rule 3: Trade Practice

The construction industry is inexact and as a result, certain tolerances have evolved. The tolerances, methods, and variances that are accepted by the construction trade normally serve as the standards that govern performance of the contractor's work. However, the contractor should never ignore the specific language of the contract. If the contract clearly establishes a specific range of allowable tolerances, for example, the contractor is held to those contract tolerances, even though a larger deviation may be generally accepted in the industry. Further, the contractor must abide by *authoritative industry practices*. The fact that non-conforming work was permitted on another job does not make it customary or industry practice. See Figure 4.14 for an explanation of *Trade Practice* contract requirements.

Rule 4: The Manifest Intent

It is virtually impossible for every detail to be unambiguously stated in the drawings or specifications. If the specifier leaves out a detail that is obviously necessary to perform a completed construction project, the contractor should furnish it. (See Figure 4.15 for a more complete explanation of the variations on Manifest Intent.)

For example, the architect may neglect to specify cover plates for the electrical outlets in an office building. Clearly, a complete office building would have cover plates for the electrical outlets; therefore, the contractor should probably price them.

The Key Test in Contract Interpretation

Read the contract reasonably.

☐ Read it from the perspective of a general contractor, not an engineer.

☐ Do not try to slant it your way.

☐ Support your position using:

    ____ *The Women In Construction Dictionary, Means Illustrated Construction Dictionary,* or a current edition of Webster's.

    ____ Industry standards, manufacturer's standards, etc.

☐ See if your Bid File (basis of your bid) supports your position. Try to show owner he is not paying for the same thing twice.

☐ Should you, as a general contractor with your expertise and experience, have known what the contract meant?

Whose risk is it if the work cannot be performed as specified?

☐ The contractor is not relieved of the obligation of financial risk just because the work is:

    ____ very difficult

    ____ very expensive

    ____ unusual (higher than normal tolerance)

☐ If the work cannot be done, such as:

    ____ two "female" interfaces specified

    ____ tolerance cannot be met (1/16th of an inch on corner bead)

    ____ performance spec cannot be met because owner criteria is wrong,
       ... then the owner has the risk.

Figure 4.13

Rule 5: Does Ordinance or Legislation Require It?

The law may impose requirements that must be fulfilled even though they are not specified in the contract documents. As an example, equipment that is installed in a project may need to meet certain standards of safety, even though these standards may not be spelled out in the contract.

Rule 6: Precedence of Contract Documents

The contract generally states the precedence to be accorded to words, terms, and drawings in the event of a conflict of information. Specifications usually take precedence over drawings, specific provisions take precedence over general provisions, etc. The contractor should be aware that the courts will make every effort possible to reconcile seemingly conflicting provisions contained in the contract documents.

Trade Practices Contract Requirements

Trade Practice...

- If the contract explicitly states a requirement (such as a certain quality or tolerance), you normally cannot go outside the contract to get another interpretation.
  - However, you can go to accepted, published standards (ACI, ASTM, etc.) to determine *accepted* tolerances.
- If the contract does not specify, or is vague, then you can go outside the contract:
  - to other jobs you have done;
  - to other jobs in the community;
  - to letters from other contractors or trade associations.
- If the job is in a remote location where labor is difficult to come by, you should always be able to look at local practice as a reference.

Non-Quantifiable Words, or No Standard for Acceptance...

- Non-quantifiable words include:
  - even
  - straight
  - true and plumb
  - non-porous
  - smooth
  - perfect weld
  - highest quality
  - quality of like luxury hotel or power plant
- What do they mean?
  In each case, go to acceptable industry standard or practice. Also, support with basis of your bid.

Figure 4.14

117

Rule 7: Construed Against the Drafter of the Document

This rule of interpretation is unfortunately often abused. It is viewed by the courts as a last resort. If after applying the first six rules, it is still impossible to arrive at a reasonable interpretation, the courts will construe the language against the party who drafted the contract language or created the drawing.

The Court of Claims has stated:

". . . The obligation to seek clarification as to patent ambiguity is inherent. However, Contractors are businessmen, and in the business of bidding on Government contracts, they are usually pressed for time and are consciously seeking to underbid a number of competitors. Consequently, they estimate only those costs which they feel the contract terms will permit the Government to insist upon by way of performance. They are obligated to bring to the Government's attention major discrepancies or errors which they detect in the specifications or drawings, or else fail to do so at their peril. But they are not expected to exercise clairvoyance in spotting hidden ambiguities in the bid documents, and they are protected if they innocently construe in their own favor an ambiguity equally susceptible to another interpretation. For . . . the basic precept is that ambiguities in contracts drawn by the Government are construed against the drafters. In the case before us, the ambiguity was subtle, not blatant; the Contractor was genuinely misled and not deliberately seeking to profit from a recognized error by the Government. "

This rule applies to private as well as government contracts. However, the contractor *must document* its interpretation!

Illustrative Case

In the case of *United States v. Spearin*, 248 E.S. 132 (1918), plans and specifications were drafted by the government to reconstruct a sewer which intersected a dry dock project site. The contractor rebuilt the sewer as required by the contract documents, and it was

---

Obvious Omissions

The Manifest Intent...

- This rule applies to absolutely obvious details that a specifier has omitted, such as:
  — face plates on electrical switches;
  — materials, accessories that are always required for installation;
  For these, the contractor is generally liable.
- This rule is not a complete "catch-all" for the owner.
- Application of the rule will vary with contract type:
  — It will be strictly construed on bid projects, especially state and municipal jobs;
  — It will be liberally construed on negotiated projects where the contractor has worked with the architect in the design phase.

Figure 4.15

accepted by the government. However, a dam (unknown to either party) was discovered in a connecting sewer —within the Navy Yard, but outside the contractor's operations. Further, due to general conditions of drainage known to the government but not to the contractor, backwaters ruptured the new sewer during a heavy rain and flooded the dry dock excavation, causing extensive damage. The contractor would not proceed with repairs unless the government paid for the damage and made safe the sewer system. In response, the government cancelled the contract. The Supreme Court of the United States, reaffirming the decision of the Court of Claims, held that when the government issues design specifications, it warrants the adequacy and accuracy of the specifications, and the contractor has the right to rely on the specifications as being accurate and correct, so that increased costs incurred by the contractor due to errors and omissions in the specifications are recoverable. As the court stated,

> *"If the Contractor is bound to build according to the plans and specifications prepared by the Owner, the Contractor will not be responsible for the consequences of defects in plans and specifications."*

This responsibility of the owner is not removed by the typical clauses requiring that builders visit the site, check the plans, and acquaint themselves with the requirements of the work. In short, the contractor can rely on the Spearin case for the thesis that data, information, and statements contained in the contract documents have the force and effect of representations, covenants or warranties, and that the owner is not protected from additional cost resulting from a breach of his representations by risk-transferring devices such as exculpatory clauses, No Damages for Delay clauses, etc.

Rule 8: The Contractor's Duty to Seek Clarification

This rule is the owner's main defense against a claim of ambiguity. During the bid phase, the contractor has a duty to inquire about major, obvious discrepancies or omissions, or drastic conflicts in the provisions. Following the Spearin Doctrine, the contractor is not, however, normally required to seek clarification of any and all ambiguities, doubts, or possible differences in interpretation during the bid phase.

Rule 9: The Conduct of the Parties to the Contract

This is the most compelling rule of contract interpretations. If the contractor fails to dispute work that he believes is out of scope and performs it without protest, he will find it very difficult, if not impossible, to come back in at a later time and claim extra work (see Figure 4.16).

In order to successfully dispute an interpretation, the contractor *must document* any discrepancy or omission as soon as it is recognized or becomes known. If the contractor is performing work under one interpretation and then decides that a new interpretation is more correct and wishes to claim the extra costs of performing the original work, he may be denied such extra costs. The parties' interpretation of a contract *before* it becomes a subject of controversy is seen by the court as having great, if not controlling influence. One of the canons of contract construction is that the interpretation given the contract by the parties to the contract during its performance demonstrates the contract's true intentions.

Most of the contracts are written by the owner or owner's representative. Consequently, they tend to favor the owner and not the contractor.

Rule 10: Document

A general contractor is responsible for all work set forth by way of the construction plans, specifications, and related documents. It is very important that the contractor ensure that all facets of the project are covered when subcontracting out work, list *all* work to be covered, designate the trade that must cover it, and ensure that that work is covered in their subcontract. *Any* and *all* interpretations *must* be written down and included with your bid documents and estimates, and subsequently during contract performance.

The Contract Document Checklist shown in Figure 3.9 is an aid for contractors to use in spotting clauses intended to transfer risk. Some of these clauses require special pricing; some do not. *All* of them require the contractor's special attention and careful management.

Summary

Contracts for construction are full of potential risks for the contractor. Those risks must be understood, priced, managed, and enforced. It is essential that the contract risks also be understood in the context of the bid/no bid decision. Some contract risks are just too immense for a given contractor to take on (for example, a small to medium-sized company bidding an energy project that has $25,000 per day liquidated damages).

The Conduct of the Parties Is the "Acid Test" of Contract Interpretation

- The most compelling rule is the interpretation the parties put on the contract by *performance* or *conduct*.
- If the *contractor*
  — fails to protest;
  — fails to give written notice;
  — after he does the work, decides the work is out of the scope and puts in a claim;
  — over a beer with the inspector, the project manager agrees with the owner's position, while his boss is claiming the work is an extra —
  then the contractor will lose.
- If the *owner* fails to object to something he should have known about, then his conduct (failure to protest) may be used against him.
- Remember, the same rules apply to the subcontractors.

Figure 4.16

The estimator must understand the risks and either price them or figure out a way to manage or avoid them. For example, in a performance specification of a high risk item, it may be wise to contract that item out.

The project manager is a risk manager all the way. He is also a contract manager—both fulfilling his company's contract obligations and ensuring that other parties (such as the owner and subcontractors) perform their duties.

All contractors should place a little sign in their offices, both home and field, with the words: "RTC: Read the Contract."

Chapter Five

Job Set-up and Planning

Chapter Five

Job Set-up and Planning

The management of a construction project involves:

- Figuring out *what* needs to be done. (We will call this *planning*.)
- Figuring out *when* it needs to be done. (We will call this *scheduling*.)
- *Causing* the performance of what needs to be done in accordance with the schedule. (We will call this *implementation* or *execution*).

This chapter is concerned with *planning*, or *what needs to be done*, at the outset of a project. Most project managers and superintendents get started on a project before either has read fully the plans and specifications, or taken the time to become familiar with the estimate. In other words, they begin *work* before they begin to *plan*. Before long, in the midst of the flurry of activity, it is not surprising that one or more items "falls through the cracks."

Planning a project is essential to its profitable and timely completion. Experience shows that the *first twenty-five percent* of a job is normally its bellwether; if the job is kicked off well, momentum established, priorities identified and managed, an effective schedule (coordinated with all affected parties) in place, submittals in process . . . then the rest of the job will normally run well.

Unfortunately, there is the reverse tendency to *under-manage* the kickoff of a project, and then intensify management as crises (which were preventable) emerge. We call this "push-pull" management, as illustrated in Figure 5.1.

One of the reasons for this failure to plan is that many contractors emphasize the acquisition of work ("I've got to have that job"), but are skimpy on the necessary project management and effort to run the job once it is acquired. Another reason for failure of the planning process is that many contractors (and their field staffs) simply do not know all that it entails, nor do they fully appreciate the cost-benefit ratio of effective planning.

For a contractor to continue successfully, planning is an absolute necessity. In fact, experience has shown that the fulcrum depicted in Figure 5.2 seems to apply to planning and productivity.

Since planning is deciding what is to be done, it is a management function, not a clerical or administrative function. It involves:

- Deciding *what* is to be done
- Deciding *who* is to do it
- Deciding *how* it is to be done
- Understanding and *managing the financial risks* as Eestablished by the *contract documents* (i.e., effective contract management).

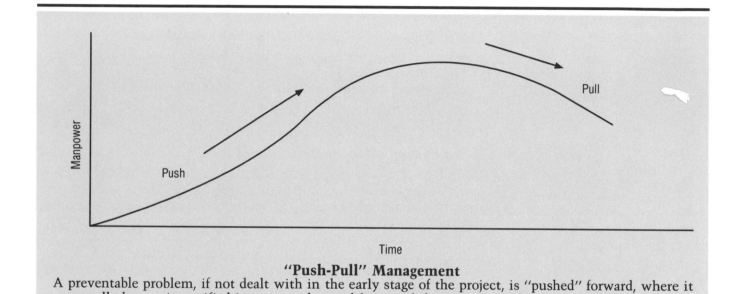

"Push-Pull" Management

A preventable problem, if not dealt with in the early stage of the project, is "pushed" forward, where it eventually has an intensified impact on the workforce and the project. At this later stage, management must "pull" the project out of a crisis.

Figure 5.1

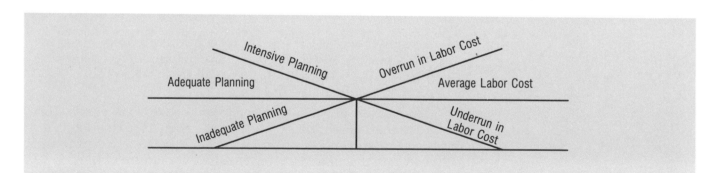

The Fulcrum Concept: The Cost Benefit Ratio of Planning

This figure illustrates how projects that are well planned at the outset, and *continuously* planned throughout the job, generally result in a labor unit cost equal to or better than the bid cost. It also shows that in general, underplanned and uncontrolled projects produce labor cost overruns.

Figure 5.2

The flow chart in Figure 5.3 can be used as a checklist for key elements which must be planned at the outset of the project.

Selecting Qualified and Adequate Management

The first step in the planning function is the selection of *qualified* and *adequate* management. Most bad jobs (not all, but most) suffer from supervisory personnel who are not fully qualified to perform the assigned duties on this particular job. The most important decision in planning, then, is *"which qualified personnel* will manage the project?" It is rare, indeed, that one can gamble successfully on unqualified, inexperienced personnel. This business is tough enough for those who are preeminently qualified and experienced. When you select competent people to

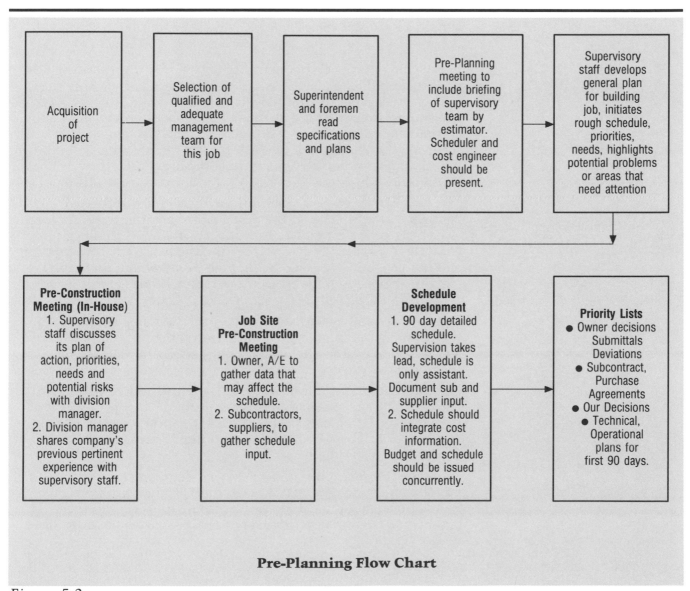

Pre-Planning Flow Chart

Figure 5.3

run a job, you are starting to minimize the risks. When you select personnel who are only marginally qualified, you are *gambling*.

There must also be *enough* personnel. Contractors tend to "go cheap" by shortchanging the project management staff, but think nothing of adding ten carpenters when the job gets into trouble—and it may well have gotten into trouble because of inadequate project pre-planning. All studies done of successful and unsuccessful projects clearly state that *an adequate and competent project management staff* is the most *singularly important criteria for a project to be run effectively*!

Familiarizing Personnel with the Contract

The second step in the flow chart is for the field supervisory personnel to become fully familiar with the plans and specifications. This is the time to begin the practice of RTC discussed in the previous chapter—**R**ead **T**he **C**ontract. In addition, on the job site, applicable and referenced standards (ASTMs, ACIs) should be made available. RTC—read the contract! How many punch lists will go away; how much quality will be improved; how many disputes can be removed if only this simple advice will be followed.

Information Transfer

In the third step, the vital information transfer from estimating to project should take place. Lack of communication remains the villain of the construction industry. In survey after survey, lack of communication is ranked as the number one problem and risk by contractors' management and field personnel. In fact, a "two company" concept has developed in many companies: a "home office company" and a "field office company," with the communication between the two sometimes quite inadequate.

The first place to attack this communication gap and "two company" syndrome is at the point that the baton transfers from one entity to the other. Remember that, in a way, that is all communication really is—transferring the baton (the information another person needs to do his/her job properly and profitably) from the person who has it to the person who needs it.

Because the estimating phase is the first bastion of essential information (and as we stated earlier, the first step in the pre-planning function), that information must be well organized and transferred to the field supervisory staff who will be managing the project.

One of the most vital meetings on any construction project is the one at which the estimator transfers to the field supervisory staff the key estimating data. Using the Information Transfer Checklist in Figure 5.4, the estimator can very quickly bring the field supervisory staff "up to speed."

Once the information transfer has been done properly, the pre-planning process is started in earnest, and on a solid foundation. That foundation (the estimate file) will aid the planners in finding a profitable, productive way to perform the work.

In step *four* (and the remaining steps) on the planning flow chart, the supervisory staff develops the plan for the project. Since project managers must *plan*, *schedule*, and *control*, then at the

Information Transfer Checklist

This checklist is to be used in assembling the information which will be transferred to the field project staff at the Pre-construction Job Coordination Meeting.

A. CONTRACT

                                                                Prepared By: Date
                                                                (Initials)

1. Contract with Owner

2. Unique Contract Provision Checklist

3. Subcontract/Purchase Contracts

4. A list of all representations to owner

5. Written Narrative—a narrative outlining any deviation from standard subcontract terms.

6. Review of status of unwritten contracts

7. Pending Changes

8. Building Permit status

9. Review billing procedures, schedule of values, retention, billing dates, project's tax status or special tax procedures, payment for stored materials, and discounts.

10. Status or narrative on negotiations leading to contract, i.e., value engineering items.

11. Review of special payment procedures for subcontractor and vendor payments

12. Overhead and fee to be charged on extra work

B. DOCUMENTS

1. Current Plan Distribution Log

2. Correspondence/telephone memos

3. Extra copies of Contract Documents (Drawings, Specifications, Addenda)

4. An index of the above information

C. ESTIMATE REVIEW

1. Any special interpretations made by estimator

2. Subcontractor/Vendor Approval form

3. Quantity Survey/Estimate

4. List of Job-Purchase items

5. Bid/No Bid Checklist

6. Bid Schedule

Figure 5.4

Information Transfer Checklist

	Prepared By: (Initials)	Date
7. Special risks (contractual or otherwise)	_____	_____
8. How the estimator visualized construction of the project	_____	_____
9. Submittal log established	_____	_____
10. Change Log established	_____	_____
11. RFI Log established	_____	_____
12. Subcontractor Log established	_____	_____

D. SCHEDULE/COST ACCOUNTING

1. Definition of the type of costs included in the cost codes	_____	_____
2. Discussion of problem areas that have been identified by subcontractors in delivery detailing, proposed alternates, etc.	_____	_____
3. Bid Schedule	_____	_____
4. Preliminary Construction Schedule	_____	_____

E. PERSONNEL

1. Subcontractor/Vendor list	_____	_____
2. Key personnel list	_____	_____

RECEIVED BY:

_____ Project Manager

_____ Date

Figure 5.4 (continued)

planning level, control must be taken into consideration. Planning cannot be simply an intellectual exercise. Planning must be action-oriented, it must be realistic, and the plan must be carried out! Project management personnel must get input from others who are committed to making the plan work. The secret of project management is to get others to *buy into the plan*—to have *ownership* of the plan, and *responsibility* and *accountability* for making the plan a reality! Thus, the plan must be a *composite of the thinking of all the parties*, and must have the *commitment of all the parties to attain it*. To many, control means monitoring the difference between the planned performance and the actual performance, and taking corrective actions. *Real control* is working to *prevent* that difference or to *improve* the plan so that corrective (and costly) action does not have to be taken. The following checklists can be useful in this vital planning function.

Project Start-up

The Project Start-up Checklist is a *memory jogger* for key supervisory personnel to review. Just like that 747 pilot reviewing his takeoff checklist to ensure that the flaps are in the correct position, the project team also must have a checklist to ensure that nothing has been overlooked. Construction projects are not forgiving of glitches—which *always* cost someone time and money. The purpose of the Project Start-up Checklist is to make sure that these steps in the flow of construction are fully considered.

The Project Start-up Checklist (shown in Figure 5.5) should be tailored to each project. For example, a federal project with its unique requirements for certified payroll, Contractor Quality Program records, etc. may require different activities than a commercial project by a private owner.

Material Handling Plan

On most projects, material management is a major factor affecting cost. The *right* material (or equipment) must be delivered to the job site at the *right time*, and *handled right* (so that there is no damage due to poor storage), *installed right*, *tested* and *checked out right*. If one of these "rights" becomes a "wrong," the installation process time is affected, and so is the cost of installation. There are risks every step of the way in material management.

The chart in Figure 5.6 shows the sequence of the material management function, and some of the key risks managers must guard against. It is helpful at the Step 4 Supervisory Planning Meeting (see Chapter 6) if the material management flow chart is reviewed by the full supervisory staff. If there is a secret to the success of the Japanese management system, it is logistics, or material management.

Quality Control

The general contractor is responsible for contract compliance for *all* of the work performed under his contract, whether performed by his own forces or subcontractors. The first step in planning for contract compliance and developing an adequate quality compliance plan is to RTC—**R**ead **T**he **C**ontract! Then it

Project Start-Up Checklist

A. Transfer Phase	Date	Initials
1. Document transfer has occurred.	_____	_____
2. Project team has read contract documents.	_____	_____
3. Contract/Estimate briefing at Pre-construction Job Coordination Meeting	_____	_____
4. Major risks identified from perspective of field staff.	_____	_____
5. Notice to Proceed issued.	_____	_____

Project start date _____

B. Pre-planning and Mobilization Phase	Date	Initials
1. Prepared schedule—began pre-planning.	_____	_____
2. Established safety and quality plans.	_____	_____
3. Required permits obtained.	_____	_____
4. Builders Risk insurance provided.	_____	_____
5. Project staff organization and responsibilities defined.	_____	_____
6. Material handling plan developed.	_____	_____
7. Staff authority communicated to owner.	_____	_____
8. Established major priorities for next 90 days and reviewed them with the owner and A/E.	_____	_____
9. Billing procedures established and schedule of values for billing submitted.	_____	_____
10. All required legal notices are posted (OSHA, etc.).	_____	_____
11. Plan to coordinate equipment suppliers and interfaces established.	_____	_____
12. Specified set of drawings to be maintained as "record set" at job site.	_____	_____
13. Defined formal change notification routing.	_____	_____

C. Job Site Administrative Phase		
1. Contract fully executed.	_____	_____
2. Owner's representative's and architect's authority clearly defined.	_____	_____
3. Schedule for periodic review meetings with owner and architect is established.	_____	_____

Figure 5.5

Project Start-Up Checklist

	Date	Initials
4. Office filing system established.	____	____
5. All subcontracts finalized.	____	____
6. Certificates of insurance obtained.	____	____
7. Subcontract bonds obtained as required.	____	____
8. Subcontractor control log set up.	____	____
9. Submittals (material/shop drawings) log completed.	____	____
10. Submittals for long lead time items expedited.	____	____
11. All major purchases committed.	____	____
12. Material commitments logged into material expediting log.	____	____
13. Requests for Information log set up and maintained.	____	____
14. Changes log set up and maintained.	____	____
15. Job diary file established and filled out daily.	____	____
16. Occurrence report log established.	____	____
17. Problem log set up.	____	____
18. Labor agreements reviewed.	____	____

Figure 5.5 (continued)

A. Estimate Phase Procedure	Personnel	Risk
Material Management **Responsibilities and Risks**		

1. Quantity Takeoff	Estimator	• Incorrect takeoff
		  a) Estimate shortfall
		  b) Estimate too high
		  c) Misread specifications

2. Quotation Request	Estimator to Buyer	• Non-spec material • Incomplete specification/listing • Outline specification • Performance specification
3. Sourcing and Selection of Vendors a. Commodity Bidders List	Buyer	• Poor quality and service • Poor delivery schedule • Weak financially • Not prequalified • Not enough vendors contacted

4. Send out RFQ's
  a. MBE, WBE, SBE
  b. Log RFQ's and maintain pending files
	Buyer	• Insufficient response time
		• Inadequate coverage
		• Incomplete RFQ date

5. Receipt of RFQ's	Buyer	• Poor vendor response • Partial response • Non-qualified response • No response • Late response

6. Evaluation of Vendor Offers	Buyer
	Estimator
		• Incorrect evaluation:
		  a) Offer conditions
		    1) Delivery time and terms
		    2) Shop drawing requirements
		    3) Payment terms
		      a) Price protection
		    4) Exceptions
		    5) Non-recognition of addenda(s)
		    6) Tax obligation (sales)
		    7) Freight requirements
		      a) Actual vs. estimated

7. Estimate	Estimator	• Incorrect spreadsheet analysis
		• Incorrect vendor selection
		  a) Non-spec
		  b) Selection by price alone
		• Long lead time items not recognized

8. Information Transfer (Successful Estimate)	Estimator to Operations	• Incomplete
		  a) Missing quotes
		  b) Missing analysis
		  c) Missing information
		    1) Cost savings
		    2) Verbal Owner-Architect/Supplier instruction/agreements
		    3) Issued addenda
		  d) Determined or assumed methods of construction…schedules not transferred

Figure 5.6

Material Management Responsibilities and Risks

B. Buyout Phase Procedure	Personnel	Risk

1. Purchase Request	Project Management (technical) Buyer (commercial)	• Untimely request (late)
		• Incomplete scope of work
		• Incorrect takeoff
		• Premature commitments
		• Not interfaced with schedule
		  a) Expediting required
		  b) Special conditions
		  c) Wrong date required
		  d) Long-lead-time items not identified or not pre-purchased
		• Wrong (non-spec) material
		  a) Wrong size, color, etc.
		• Incorrect cost code
		• Inadequate review by buyer
		• Arbitrary vendor change by the project engineer
		• Improper evaluation of storage vs. bulk buying and shipping, etc.

2. Purchase Order
  • Logs out P.O. document
  • Distributes P.O.
    a) Vendor
    b) Accounting
    c) Purchasing
      1) Traffic
      2) Expediting
      3) Buyer file
    d) Operations
	Buyer	• Input error
		• Incomplete description
		• Incorrect pricing
		• Order not accepted by Vendor
		• Vendor changes conditions
		• Expired quotes
		• Verbal quotes different from confirmation quotes (written) after bid date
		• Buyer lack of negotiation
		• Shipment time not confirmed

3. Submittals	Project Management	• Late submittal requests • Late submittal approvals • Changes by owner/architect • Incomplete

4. Expediting	Project Manager (Expeditor)	• Incorrect requirement dates
		• Not coordinated with schedule
		• Late requests
		• Change in schedule
		  a) Air shipment
		  b) Price increases (expired quote)
		  c) Multiple shipments
		    1) Partial air freight
		    2) Partial normal routing

5. Delivery	Buyer Project Engineer	• F.O.B. Point
		• Transportation from dock or vendor's warehouse to job site
		• Unloading equipment availability/unavailability
		• Preplanning of "hook" time
		• Lost man-hours of productivity
		• Receipt and timely inspection of material
		  a) Inventory inspection
		    • Shortage/overage • As ordered
		                 • At plant prior to shipping
		    • Quality • Lack of enforcement
		    • Damage of P.O. Agreements
		  b) Payment terms
		  c) Delivery terms, etc.

Figure 5.6 (continued)

135

Material Management Responsibilities and Risks		
Buyout Phase Procedure *(continued)*	**Personnel**	**Risk**
6. Invoice Payments	Project Engineer Accounting Project Manager	• Late approval • Loss of cash discounts • Vendors not shipping until accounts are clear • Vendor's unwillingness to quote if payments are continually late • Incorrect cost coding

C. Installation Phase Procedure
   1. Storage
	Job Site
	Personnel
		Preplanning
		• Available area
		• Closeness to job need
		• Suitability of storage area
		Environmental Considerations
		• Weather
		• Suitability of ground
		• Access
		• Weathertightness of storage facility
		Warehousing
		• Inventory Log
		  • Sign out for material
		  • Maintain current status
		    • Prevent shortages/theft when needed
		Theft
		  • Maintain security
		    • Guard • Fence • Shack
		  • Maintain log of missing material
		Breakage
		• Improper handling

2. Material Movement	Job Site
	Personnel
		Premature Delivery
		• Too long storage period
		  1) Deterioration 4) Change orders
		  2) Theft 5) Storage requirements
		  3) Owner 6) Inefficiencies of moving
		    payment terms materials more than required
		• Problems in locating material
		• Safety

3. Job Environment	Job Site
	Personnel
		Accessibility to work - not impeded by material (new and discarded)
		Distribution
		1) Material available when and where required to maintain productivity
		  a) correct material c) undamaged
		  b) correct number
		Security • Lighting • Parking
		Safety • Power • Trade/Subs' Trailers
		Toilet/Sanitary Facilities
		2) Man-hours lost waiting for or replacing wrongly distributed material
		3) Untimely distribution

Figure 5.6 (continued)

Material Management Responsibilities and Risks		
Installation Phase Procedure (continued)	**Personnel**	**Risk**

4. Installation	Job Site Personnel	• Waste Control
		• Shop drawing approved
		• Correct tools
		• Equipment availability
		• Current Material Status Report
		  • Interfaced with:
		    • P.O.'s
		    • Change orders
		    • Phases
		    • Outstanding orders
		• Recognition of "special" or unique installation techniques
		    • Manufacturers' Representatives
		    • Spare parts
		• Obtaining lien releases from suppliers (prime and subcontractor)
		• Safety

Figure 5.6 (continued)

becomes important to *communicate* all requirements to each and every member of the supervisory personnel with *accountability* for quality control and contract compliance.

Quality is not to be taken lightly. Owners and users demand the quality specified and will not hesitate to sue the contractor whose performance is substandard or causes damage, or to use the contractor's poor or substandard quality as a defense to an otherwise valid contractor's claim. Poor quality adversely affects not only productivity and unit cost, but also work force morale. Contractors who fail to plan for quality and ensure that it is achieved have a limited life in the construction business. Quality planning almost always improves the scheduling process. And, of course, safety planning is inherent in quality planning. The Quality Control Checklist in Figure 5.7 is a guide for establishing a quality plan, and maintaining the required level of quality.

Cost Accounting

The cost accounting system is an essential management tool. Cost reports that are historical, inaccurate, or difficult to use should be junked. The field *must have* timely data so it knows

Quality Control Checklist

An individual on the office staff should be responsible for identifying the contractual quality requirement. The Project Superintendent will be responsible for implementing this quality control.

A. QUALITY PLAN   Initials   Date

1. Review contract requirements for Quality Plan   _____ _____

2. Identify special or unique quality requirements   _____ _____

3. Develop Quality Requirements Plan (Input from subs)   _____ _____

4. Review Quality Requirements Plan with supervisors   _____ _____

5. Review subcontract provisions for contract options if subs' work is non-compliant (notice to cure, payment withholding, etc.)   _____ _____

6. Comments _____

B. MANAGING QUALITY

1. Set Standards

   a. Before each new work activity, review quality requirements with supervisory personnel.

   b. Establish aesthetic "finish standards" for such trades as painting, etc.

   c. Maintain all test data results.

   d. Adhere to quality standards after the beginning of the project. Don't wait for complaints. Run your own punch lists!

2. Schedule Inspections

3. Manage Mistakes

   a. Correct them quickly.

   b. Learn from them.

   c. Don't Repeat them!

   (Note: Contractor is responsible for quality. Do not abdicate to the architect.)

4. In cases of dispute regarding tolerances or non-compliance:

   a. Double check contract requirements.

   b. Check bid file for basis of estimate and interpretations, or modifications.

   c. Obtain reliable industry standard.

   d. If necessary, bring in consultant *promptly*.

   e. If necessary, bring in legal services *promptly*.

Figure 5.7

Quality Control Checklist

4. Document inspection results (discussions with inspectors, etc.)

   a. Schedule inspections

   b. Concrete pours

   c. Concealed inspections

   d. List types of inspections needed:

_____

_____

_____

_____

_____

_____

Figure 5.7 (continued)

where it is going and has the time to take corrective actions. The field "cheats" by mis-reporting or failing to report cost information. It destroys any opportunity that top management may have to take corrective action. Furthermore, inaccurately reported cost information may be worthless in a claim or litigation situation. The Cost Accounting Checklist in Figure 5.8 lists criteria that should be considered in setting up a cost accounting system, and some basic requirements for select reports. It also lists some fundamental standards for implementing a cost accounting system.

Project Control and Record Keeping

When the project finally begins, the pre-planning is put to the test and plans are adjusted as needed. There must be methods in place to verify that events have occurred or will occur. These methods protect the contractor against losing control of the job. The paper or electronic trail that records important events on a construction project can be as essential to the well being of a project as the pre-planning effort.

Logs

A key element in the paperwork that supports project control is logs. The most familiar type of log, which is used on even the smallest of jobs, is the superintendent's daily log, used to note the weather, significant events, etc., at the end of each day. However, a formal method of tracking details is necessary for good project control.

Figures 5.9-5.12 are types of logs that should be used on each project. They are the Change Order log, the Request For Information log, the Material Status Report, and the Shop Drawing/ Submittal Log.

Managing Changes

Changes are inevitable, despite the best laid plans and promises to the contrary. The wise general contractor accepts this fact and plans for changes, but does not implement them or spend any money on them without proper authority and documentation.

Changes are a problem in the construction industry, largely because they can create confusion and, consequently, delays. Timely and properly documented communication is the key to solving this problem. The contractor must first have a clear understanding of the current status of the change. Secondly, that understanding must be communicated to all concerned parties inside and outside the contractor's organization.

All personnel on the contractor's office staff at the project should be responsible for managing contract changes and the associated communication requirements. All should be responsible for identifying changed conditions. These requirements mean that project personnel must recognize items that represent an extra cost to the contractor, and for which the owner should pay. Figure 5.13 categorizes changes due to conditions that differ from those described in the contract documents versus changes that are time-related. Compensable and noncompensable delays are also identified.

The Managing Changes Checklist in Figure 5.14 provides the contractor with both a strategy for change management and a

Criteria for Effective Cost Accounting/Reporting System

A. **General Considerations**
1. Consider historic cost codes; is all historical data relevant?
2. Consider how field staff will use accounting system.
   a. Do not make it too detailed for the field.
   b. Include summary codes for project overview.
   c. Consider organizing report into work breakdown structure (WBS) vs. pure activities.
   d. Show cost trends to pinpoint areas that need attention.
3. Consider upper management.
   a. Make reports general enough to convey trends of project without being lost in detail.
   b. Provide early warning indicators.
4. Organize report such that it may be integrated with schedule activities.
5. Discuss the cost code structure with field staff and management so they understand the reports.

B. **Report Criteria**
1. Use a minimum of paperwork to tell the story.
2. Consider processing concise reports on site. In this way, the information will be more timely, as it will not involve the long turn-around time of central processing.
3. Consider organizing data in division summaries, rather than scattered details.
4. Consider showing historical job site trends:
   — will allow for more accurate projections;
   — will help the project team spot inconsistencies.
5. Consider user priorities.
   a. Estimating department needs final unit prices and/or productivity factors for historical data.
   b. Field needs to see trends at the summary level.
   c. Upper management needs basic information in order to project costs.
6. Use graphs to illustrate trends.
7. Integrate with schedule.
   a. Consider earned value analysis.
   b. Consider schedule "crash" analysis.

C. **Implementation**
1. Be sure field staff understands use and benefit of cost reports.
2. Set up quantity reporting system early in the job.
   a. Break quantities down into units (e.g., tons of rebar per column and percentage for pre-tie).
   b. Convey the need for reasonably accurate quantities.
   c. Detail of reported quantities should not exceed the level that will be checked.
3. Establish a method for handling corrective work (special code or divide into standard codes).
4. Establish a method for handling changes.
5. Cost engineer should control cost accounting throughout the project duration.
6. Perform realistic cost-to-complete exercises in order to know the true project status.

Figure 5.8

Change Order Log

Project _____ Job No. _____

Request for Proposal			Proposal							Authorization		
					To Owner		Days	Approval				
No.	Date	Description	RFI No.	To Subs (list)	From Subs (list)	Date	$ Amount	Elapsed	Date	$ Amount	Notice to Proceed	Change Order No.

Figure 5.9

Request For Information (RFI) Log

Project _____ Architect _____ Owner _____

Request Number	Date of Request	Description	Specification Drawing No.	Disposition	Disposition Date	Days Elapsed	Effect*

*Record Effect on Occurrence Report

Figure 5.10

Material Status Report Log

Project _____ Job No. _____

Material Item	Specification			Delivery Date		Quantity		Submittal Date		Approval Date	
	Div.	Sec.	Pg.	Promised	Actual	Ordered	Received	Expected	Actual	Expected	Actual

Figure 5.11

Shop Drawing/Submittal Log

Project _____ Job No. _____

Item Description	Specification			Drawing/Sample/ Cut Number	Submittal Dates		Disposi- tion*	Returned to Sub
	Div.	Sec.	Page		To Us	To A/E		

*A = approved; AAN = approved as noted; R = rejected; RR = revise & resubmit
NOTE: Alternating lines are dashed to allow for resubmittals of the same item.

Figure 5.12

format for control of change procedures. Documented communication to all parties affected by the change is the key to managing changes correctly.

Differing site conditions is a special case of changes which occurs frequently and can have dramatic effects on the project. For this reason, a separate Differing Site Conditions Checklist is provided in Figure 5.15.

Job Site Filing System

The job site filing system is essential not only for the duration of the job, but often years later (if there are claims, litigation, etc.). Figure 5.16 is an organizational overview of a filing system which has proved to be extremely effective and which provides for the retrieval of information. Figure 5.17 is a list of the specific documents to be included within the divisions. Figure 5.18 is a sample list of abbreviations that are to be used in the system. The job site filing system description (including abbreviations list) should be distributed to all who will be using the system to ensure proper and consistent reports and reviews.

Early Indicator System

Finally, does the plan work? What's *happening* on this project anyway? There must be an effective, established method for gathering information about a project, instead of simply a job walk-through by the president asking why the concrete placement is overrunning the budget. (Normally, the president is really not interested in *why* concrete is overrunning—he just wants it to *stop*!) If top management can learn *early* what is happening on the job, they have a good chance to take aggressive action which *can* stop those overruns.

Identifying Changes and Delays	
Changes	
Contract Documents	**Time-Related**
Defective design	Acceleration
Discrepancy in plans	Constructive Acceleration
Added, altered, or deleted work	Suspension of work
Differing site conditions	Out-of-sequence work
	Overtime
	Stand-by
Delays	
Compensable	**Non-Compensable**
Owner's act or neglect	Adverse weather conditions
A/E's act or neglect	(unless contract indicates otherwise)
Other prime's act or neglect	Casualties beyond control (e.g.,
Changes to work	vandalism, accidents, fires, floods, riots)
Owner-authorized delay	Acts of God

Figure 5.13

Managing Changes Checklist

IDENTIFY CHANGE

☐ Sources

    ___ Field Directives

    ___ Marked up Shop Drawings

    ___ Correspondence

    ___ Formal Change Request

    ___ Field Occurrences

        ___ Differing Site Conditions

        ___ Suspension of Work

        ___ Overtime or Stand-by

☐ Documentation

    ___ Changes Log

    ___ Separate cost codes

    ___ Identify as change in other logs (i.e., comments in RFI Log)

☐ Notification

    ___ Written Directive from Owner

    ___ Written Notice to Owner

    ___ Written Directive to Subs

    ___ Written Directive to Field Personnel

    ___ Reserve Your Rights

☐ No Ambiguity in any Written Notice/Directive

    ___ Proceed or not proceed

    ___ Price before proceeding or price while proceeding

    ___ Estimate of time & material

    ___ Does person directing the change have authority?

___ Time extension/delay

___ Watch out for:

    ___ "Owner would like"

    ___ "Architect Request"

    ___ "We think the Owner wants"

___ Precise scope of work delineated

___ Change Order will follow

___ Amount of overhead & fee

___ Payment provisions

DOCUMENTATION AND CONTROL OF CHANGES

☐ Pricing

    ___ Stay on top of pricing effort

        ___ Get help if behind

        ___ Enforce timely receipt of Sub/Vendor quotes

        ___ Do not let dime hold up a dollar (i.e., $40,000 quote held up by $150 vendor quote)

    ___ Pricing entire change

        ___ Impact schedule to see effect

        ___ Price all items

        ___ Keep direct, indirect and impact cost separate

    ___ Press Owner to accept or reject pricing

☐ Communication

    ___ Press Architect for revised drawings

    ___ Be sure Architect and Owner are clearly notified of:

        ___ What you are doing

        ___ Compensation you expect

Figure 5.14

Managing Changes Checklist

    ____ Give Owner information needed
           to make timely decision

    ____ Be sure Sub/Vendors are clearly
           notified of:

        ____ What they are to do

        ____ Compensation they will receive

    Be sure field personnel understand exactly what they are to do

☐ Completion of Changes

    ____ Press Owner for a Formal Change Order

    ____ Execute Sub/Vendor Change Order as soon as possible

    ____ Include change on pay application as soon as possible

    ____ Integrate change into cost accounting as soon as possible

    ____ Review Changes Log weekly

    ____ Do not let outstanding changes "pile up"

    ____ Update as-built drawings

Figure 5.14 (continued)

Differing Site Conditions Checklist

☐ READ the differing site condition clause of the contract. Follow its procedural requirements.

☐ Give IMMEDIATE and WRITTEN notification to owner and ask for direction!

☐ Take PICTURES.

☐ Get Owner's representative to site immediately!

☐ If in doubt, get soils engineer or other consultant to site immediately.

☐ Set up SEPARATE COST CODE and track direct cost associated with the extra work. Daily diaries show extra work being done each day, the crew, equipment, overtime, etc.

☐ IMPACT SCHEDULE to determine what additional activities are being affected.

☐ Attempt to develop an analysis of additional time required and schedule on an optimum rather than crash basis.

☐ If no or inadequate time is permitted by Owner, DOCUMENT EXTRA COST according to procedure on acceleration.

☐ Remember, most differing site condition claims arise during crucial first 25% of job; this is where a change in sequence or momentum can have a devastating effect on the job, so the changed condition must be really well managed!

☐ Conditions which differ materially from those indicated in the contract.

   ___ Show what the actual conditions are.

   ___ Show what conditions were indicated by the plans, specs and other contract documents.

   ___ Put Owner on notice of changed conditions immediately!

☐ Unknown physical conditions of unusual nature differing materially from those *ordinarily encountered.*

   ___ Show that the contractor should not have reasonably anticipated the condition.

   ___ Show that the pre-bid site investigation was adequate and reasonable.

   ___ Show that the bid anticipated with regard to this condition and how it was "covered."

☐ Bring in an expert early to identify differing conditions and propose possible alternatives.

☐ Prepare list of options for Owner. Get Owner's direction.

Figure 5.15

The Early Indicator Checklist (Figure 5.19) becomes an important *control* tool for all the jobs. It provides a method for overseeing all the projects to enable early warning *signals* to be transformed into early management actions!

The project team should *plan* to gather the appropriate information which will provide the necessary early warning signals. With early warning signals, action can be taken while it can still do some good, whereas later information merely confirms that something bad has happened, but it is too late to do anything about it.

Summary

Again, the basic planning steps are:

Bid/No Bid Decision
- Is schedule realistic?
- Are resources available?
- Do we have supervision available?

Estimate
- Pricing schedule
- Time sensitive pricing
- Production flow and work sequence
- Complexity and its effect on time
- Contract provisions (such as "no damages" clause)

Job Setup
- Commitment to first 25% of the job
- Ninety day plan
- Logistics plan
- Budget correlated to production rates
- Priority items identified and managed
- Tie subs/suppliers contractually to schedule
- Pre-planning tools:
    Lift drawings
    Models
    Installation drawings
- Decision making
- Technical pre-planning
- Crew planning
- Safety
- Quality control
- Close-out
- Space
- Sequence
- Restraints

If there is one principle that cannot be overemphasized, it is that scheduling is a management function—a proactive, forecasting and control function. The following sections address each component in managing the scheduling process.

Job Site Document Filing System
Organizational Overview

- ☐ Preconstruction:
  All items regarding the setting up of the job site, type of structure, etc., before any subcontractor is physically on the site.

- ☐ Bid Documents:
  All of the information we need in order to prepare a bid, and all the papers regarding the bid.

- ☐ Contractor Documents:
  These should be specific contract papers. Also included: all items relating to CHANGES in the contract including work added, taken away, changed, and any proposed changes.

- ☐ Subcontractor Contract Documents:
  Concrete, Electrical, Mechanical, Rebar—all of the trades needed to build the project. Also, the material suppliers: who provides wood, scaffolding, concrete, rebar, tools, etc.?

- ☐ Schedules:
  The general contractor usually provides the main schedule with updates. But the subs, architect, and owner often have their own schedules to document their actual progress on the job.

- ☐ Claims/Backcharges:
  These are claims that are put together on the job site—not in our offices.

- ☐ Meeting Minutes/Daily Reports:
  This includes meetings that were held with two or more people, where the meeting was documented (not just a memo); any meeting that was held regularly; a meeting that was held to decide an issue. Daily Reports are included because they are a regular listing of the day's activities.

- ☐ Correspondence:

  a. Any and all correspondence—letters, memos—to and from two parties.

  b. Transmittals are included because they are a form that is filled out to show that something was sent from one party to another (drawings, blueprints, pay estimates, etc.).

  c. RFI (Request For Information); one party requests from another an answer to a question, clarification of a drawing, etc.

  d. Occurrence Report: a report stating that something out of the ordinary happened.

  (Note: For b., c., and d.—usually specific forms are used; copies of any of these may be scattered throughout the job files.)

Figure 5.16

Job Site Document Filing System
Organizational Overview

☐ Quality Control:

   a. Tests are performed on concrete and other aspects of a job; there is usually documentation of the results.

   b. Sometimes a third party is called in to evaluate some aspect of the job.

   c. Punch lists are done at the end of a job to make sure all items are complete and done to the owner's, A/E's, or contractor's satisfaction.

☐ Shop Drawings and Submittals:
   Shop Drawings: Drawings of items to be fabricated and installed by a particular subcontractor (i.e., mechanical, electrical, rebar, elevators, etc.). Submittals: Brochures for specific items to be purchased by a subcontractor or supplier and installed on the project.

☐ Pay Estimates:
   Usually a month-by-month accounting of the money due to the subs from the contractor, to the contractor from the owner, etc.

☐ Job Cost Reports:
   Computer print-outs showing the dollars spent on labor, material, or equipment, either weekly or monthly, broken down into general or specific categories.

☐ Photos:
   Job site photos: Sometimes taken through the duration of the job, sometimes only during one portion of the job, sometimes for specific reasons (like problems, damage, claims).

☐ Miscellaneous:
   Employment records, etc. Items that absolutely do not fit into any other category. BUT—NOTHING STAYS FILED HERE UNLESS SPECIFICALLY AUTHORIZED BY THE ENGINEER IN CHARGE OF THE CASE.

☐ Legal Files

Figure 5.16 (continued)

Job Site Document Filing System
Categories and Possible Items to be Included in Each Category

- ☐ Pre-construction Documents
  - ____ Architect's contract
  - ____ Engineering studies
  - ____ Soils information
  - ____ Utilities information
  - ____ Preliminary schedules
  - ____ Architect's estimate
  - ____ Other
- ☐ Bid Documents
  - ____ Plans
  - ____ Specifications
  - ____ Contract documents
  - ____ General terms and conditions
  - ____ Special terms and conditions
  - ____ Pre-bid meetings
  - ____ Bids, abstract of bids
  - ____ Bid verification
  - ____ Contractor's estimate
  - ____ Subcontractor's estimate
  - ____ Site inspection notes
  - ____ Bid schedules, subcontractor quotes
  - ____ Quantity takeoffs, productivity analysis, etc.
- ☐ Contract Documents
  - General Contractor
  - ____ Letter of Intent
  - ____ Contract with owner
  - ____ Contract with subcontractors
  - ____ Notice to Proceed
  - ____ Change orders: Log—listed by date

Figure 5.17

Job Site Document Filing System
Categories and Possible Items to be Included in Each Category

    ____ Extra work orders

    ____ Proposals: Log—listed by date

    ____ Request for Proposals (RFP)—listed by date

☐ Owner Documents

    ____ Letter of Intent

    ____ Contract with architect

    ____ Contract with general contractor

    ____ Notice to Proceed

    ____ Change orders: Log—listed by date

    ____ Extra work orders: Log—listed by date

    ____ Proposals: Log—listed by date

    ____ Request for Proposals (RFP)—listed by date

☐ Architect/Engineer Documents

    ____ Contract with owner

    ____ Notice to Proceed

    ____ Change orders: Log—listed by date

    ____ Extra work orders

    ____ Proposals: Log—listed by date

    ____ Request for Proposals (RFP)—listed by date

☐ Subcontractor Documents

    ____ Contract with general contractor

    ____ Notice to Proceed

    ____ Change orders: Log—listed by date

    ____ Extra work orders

    ____ Proposals: Log—listed by date

    ____ Request for Proposals (RFP)—listed by date

Figure 5.17 (continued)

Job Site Document Filing System
Categories and Possible Items to be Included in Each Category

☐ Project Schedules and Expediting Documentation
  ____ Contract schedule
  ____ General contractor's schedule and updates
  ____ Shop drawings and Submittal Log:
    ____ Contractor's
    ____ Owner's
    ____ Architect's
    ____ Subcontractor's
  ____ Owner's Schedule
    ____ Owner's Analysis of Schedule
  ____ A/E's Schedule
    ____ A/E's Analysis of Schedule

☐ Claims/Backcharges
  ____ Contractor
  ____ Subcontractor
  ____ Owner
  ____ A/E
    ____ Litigation record
    ____ Other

☐ Meeting Minutes/Daily Reports
  ____ Owner
  ____ Architect
  ____ Pre-Construction Conference
  ____ Project management meetings
  ____ General contractor meetings
  ____ Subcontractor meetings
  ____ Negotiation
  ____ Diaries, Daily Reports, Field Reports
  ____ Other

Figure 5.17 (continued)

Job Site Document Filing System
Categories and Possible Items to be Included in Each Category

☐ Correspondence

    _____ Contractor to owner

    _____ Owner to contractor

    _____ Contractor to subcontractor

    _____ Subcontractor to contractor

    _____ Contractor to suppliers

    _____ Suppliers to contractor

    _____ Owner to architect

    _____ Architect to owner

    _____ Architect to third parties

    _____ Third parties to architect

    _____ Subcontractor to architect

    _____ Architect to subcontractor

    _____ Subcontractor to suppliers

    _____ Suppliers to subcontractor

    _____ Other: _____ Transmittals

                  _____ Request for Information: Log, File

                  _____ Occurrence Reports: Log, File

☐ Quality Control

    _____ Soil test results

    _____ Concrete test results

    _____ Steel test results

    _____ Letters of non-compliance

    _____ Contractor's response to non-compliance

    _____ Third party evaluations

    _____ Inspections

    _____ Punch list

Figure 5.17 (continued)

Job Site Document Filing System
Categories and Possible Items to be Included in Each Category

- ☐ Shop drawings and submittals
- ☐ Pay estimates
- ☐ Job Cost Reports
  - ___ Reports (usually monthly, include all cost codes)
  - ___ Labor (usually weekly, include only labor accounts but in more detail)
  - ___ Material (usually monthly)
  - ___ Equipment (same as labor)
- ☐ Photos
- ☐ Miscellaneous
- ☐ Legal Files

Figure 5.17 (continued)

List of Abbreviations

The job site filing system description (distributed to all who will be using the system) should include a list of abbreviations, such as shown below.

Abbreviation	Document Type	Abbreviation	Document Type
ADD	Addendum to Contract	MTG	Meeting minutes
ASB	As-built drawings	NTP	Notice to Proceed
BID	Bid Proposal and Documents	OR	Occurrence Report (Notification of a problem)
CHK	Checks		
CNT	Contract	PAY	Payment application (monthly draw request)
CO	Change Order/Modification		
CSC	Certificate of Substantial Completion	PL	Punch list
		PMT	Permit
DR	Daily Report (Daily Diary, Foreman's Diary, Personal Diary)	PO	Purchase order
		PPL	Proposal
DWG	Drawings, plan sheets	RFI	Request for Information
EST	Estimate	RPT	Report (field reports)
FA	Force Account/Extra Work Authorization	SCH	Schedule
		SKT	Sketch (detail drawing)
INV	Invoice	SML	Submittal
IR	Inspection Report	SPC	Specifications
JCR	Job Cost Report	SUB	Subcontract
JI	Job Instructions/Architect's Supplemental Instructions	TCN	Telephone Conference Notes
		TEL	Telegram/Mailgram
LCR	Labor Cost Report	TER	Time Extension Request
LOG	Shop Drawing Log/Submittal Log	TLX	Telex
		TML	Transmittal form
LTR	Letter	TR	Test Report
MEM	Memo	TRG	Transaction Register
NOT	Notes: to file, handwritten, etc.		

Figure 5.18

Early Indicator Project Status Checklist

This checklist should be reviewed and updated on a regular basis throughout the life of the project.

OUTLINE

A. Estimate/Budget
B. Change Order Status
C. Clarifications
D. Schedule
E. Cost
F. Progress Billings
G. Subcontractors
H. Other
I. Management Reports

A. Estimate/Budget

		Yes	No	Date
1.	Has the construction budget been developed/balanced?	___	___	___
2.	Have change orders been integrated into construction budget?	___	___	___
3.	Have any significant estimate busts been identified?	___	___	___

Comments: _____

_____

B. Change Order Status

		Yes	No	Date
1.	Are the changes processed to date reasonably current?	___	___	___
2.	Are we receiving excessive number of changes?	___	___	___
3.	Do we maintain a comprehensive change log?	___	___	___
4.	Approximate number of change proposals outstanding —i.e., any not yet included in change orders? _____			___
5.	Estimated value of outstanding changes:			

GC: _____   _____

Subs: _____   _____

		Yes	No	Date
6.	Do our change proposals include both impact *costs* and *time*?	___	___	___
	Indirect (extended job overhead)?	___	___	___
	Direct (inefficiencies, etc.)?	___	___	___
7.	If not, are we reserving our right to claim for these at a future date?	___	___	___
8.	Are we submitting our change proposals and supporting documents on time per contract requirements?	___	___	___

Figure 5.19

Early Indicator Project Status Checklist

	Yes	No	Date
9. Is the owner reasonable in responding to changes?	___	___	___
10. Does the owner/architect process change orders in a timely fashion?	___	___	
11. Do change orders require the lender's approval?	___	___	
12. Are there limitations imposed on change order amounts (by the owner, lender, or other agency)?	___	___	
13. Do we require written a directive from the owner before proceeding with a change?	___	___	

Comments: _____

C. Clarifications

	Yes	No	Date
1. Do we maintain an RFI and Problem Log to track problems and/or questions that arise?	___	___	
2. Does the owner (architect/engineer) take action and/or respond in a timely manner to resolve the issues?	___	___	
3. Are there outstanding RFI's or problems impacting the work?	___	___	

Comments: _____

D. Schedule

	Yes	No	Date
1. Has the schedule been sent to the owner and subcontractors?	___	___	___
2. Does the updated schedule incorporate change/delay impact?	___	___	___
3. When was the schedule updated?			___
4. Is the project on schedule in accordance with the updated schedule?	___	___	___
5. Do we have adequate manpower and equipment to maintain (or get back on) schedule?			
GC:	___	___	___
Subs:	___	___	___
6. Are there any problems with anticipated late deliveries of critical materials?			
GC:	___	___	___
Subs:	___	___	___

Figure 5.19 (continued)

Early Indicator Project Status Checklist

	Yes	No	Date

7. Will there be any material delivery problems resulting from the impact of changes (change orders, proposals, etc.)? _____ _____ _____

    GC: _____ _____ _____

    Subs: _____ _____ _____

8. Has the Owner been advised of any delays affecting anticipated turnover dates or Owner coordination items? _____ _____ _____

9. Is the project experiencing delays/interference from the Owner or his agents? _____ _____ _____

10. If so, have we notified him on a timely basis and in accordance with any contract requirements? _____ _____ _____

    Comments: _____

_____

E. Costs

1. Does the Cost Report indicate any significant variances in major work accounts? _____ _____ _____

2. Does the Cost Report show any definite week to week trend at the summary level? _____ _____ _____

3. Are we performing work on changes prior to receiving an approved Change Order? _____ _____ _____

    If so, what is the approximate cost incurred to date on these unapproved changes? _____

        Contractor: _____

        Subs: _____

    Comments: _____

_____

F. Progress Billings

1. Are progress billings being regularly submitted as per contract? _____ _____ _____

2. Are our billings current? _____ _____ _____

    How does the progress billings percent complete compare to the percent complete as shown on the cost report? _____

3. Does this project show Costs in Excess of Billings? _____ _____ _____

4. Do we bill for work performed on unapproved changes? _____ _____ _____

Figure 5.19 (continued)

Early Indicator Project Status Checklist

	Yes	No	Date

5. Are progress billings generally approved as submitted, (i.e., reasonable adjustments only)? _____ _____ _____

6. Do we receive our payments on time as per contract? _____ _____ _____

Comments: _____

G. Subcontractors

1. Do we have any problem subcontractors? _____ _____ _____

2. Do we have a good relationship with each of our subcontractors? _____ _____ _____

3. Do we hold regular subcontractor meetings? _____ _____ _____

4. Are any subs unable or unwilling to provide adequate manpower to meet our production schedule? _____ _____ _____

5. Any subs unable or unwilling to provide acceptable quality workmanship? _____ _____ _____

6. Any subs requesting/requiring assistance in material management, expediting? _____ _____ _____

7. Are sub pay requests realistic in view of work performed? _____ _____ _____

8. Are any subcontractors requesting advance payment, joint checks, or other special payment arrangements? _____ _____ _____

9. Are we obtaining lien releases from all major suppliers to our subcontractors as per subcontract conditions? _____ _____ _____

10. Do we have current certificates of insurance on file for each subcontractor on the job? _____ _____ _____

11. Are all communications from the subcontractors being channeled through GC and not sent directly to the Owner/Architect? _____ _____ _____

Comments: _____

Figure 5.19 (continued)

Early Indicator Project Status Checklist

H. Other

    1. How are the relationships between:

                                                   Poor – Excellent
                                                     (1 – 10)

GC and the Owner? _____

GC and the Architect? _____

The Owner and the Architect? _____

Subcontractors and Owner/Architect? _____

	Yes	No	Date
2. Are all logs being maintained in accordance with company procedures?	_____	_____	_____
3. Does GC's project staff have regular (weekly) meetings with the owner and/or the architect?	_____	_____	_____
4. Are we experiencing any unusual or unreasonable inspection requirements from the Owner (or his architect, engineer, etc.) or any government inspectors?	_____	_____	_____
5. Have we encountered any unexpected unusual conditions which may impact the project?	_____	_____	_____
6. Have we deviated from the plans and/or specifications?	_____	_____	_____
If so, have we received written approval from the Owner and/or Architect?	_____	_____	_____
7. Are daily diaries being maintained?	_____	_____	_____

Comments: _____

_____

I. Items to be Included in Management Reports

  ☐ Continuing Problems

    ____ Continuing problems from last month which were resolved _____

    ____ Continuing problems from last month not resolved _____

    ____ Continuing problems to be added _____

    (Note: It must be emphasized that there is danger in trying to hide a problem.)

  ☐ Project Claims

    ____ Claim (by owner, subcontractors, etc.)? _____

    ____ Who is responsible for resolving? _____

Figure 5.19 (continued)

Early Indicator Project Status Checklist

____ Schedule for resolution _____

____ Potential or Actual Claim _____

____ Amount of Claim (+/–) _____

____ Nature of conflict _____

____ Probability of recovery (+/–) _____

☐ Profit/Cash Flow

____ Bit profit _____

____ Projected profit _____

____ Percent complete _____

____ Major categories of loss _____

____ Recovery plan _____

____ Who is responsible? _____

____ Payment problems (owner or sub/vendor) _____

    ____ Payment requested date _____

    ____ Amount (+/–) _____

    ____ Reason for withholding _____

    ____ Schedule for resolution _____

    ____ Responsible person _____

____ Amount of retainage _____

____ Expected release of retainage _____

☐ Unauthorized Work _____

____ Work being performed _____

____ Approximate amount _____

____ Reason for performing _____

____ Status of resolution _____

____ Schedule for resolution _____

☐ Schedule Status

____ Original contract substantial completion date _____

____ Original contract final completion date _____

____ Time extensions in contract _____

____ Time extensions not in contract _____

Figure 5.19 (continued)

Early Indicator Project Status Checklist

___ Projected substantial completion date _____

___ Projected final completion date _____

___ Percentage complete according to pay application _____

___ Percentage complete according to cost accounting _____

___ Date of last schedule update _____

☐ Changes

___ Changes identified (number and dollar amount) _____

___ Changes issued by owner _____

___ Changes priced _____

___ Changes proceeding on with no directive _____

___ Changes proceeding on with written directive _____

___ Changes in an executed change order _____

Figure 5.19 (continued)

Chapter Six

Management and Legal Implications

Chapter Six

Scheduling: Management and Legal Implications

The General Contractor's Role

General contractors are hired in large part to purchase services, schedule, and coordinate a project. Theoretically, this is their expertise. In fact, while tools such as CPM are available, *use* of those tools is not very advanced in the construction industry. Unfortunately, scheduling is too frequently treated as little more than an exercise on paper that must be done in order to fulfill a contractual agreement. CPM's are often what we call "dead trees"—inert paper rather than effective and dynamic management tools.

Proper scheduling is simply a proactive, cost-effective management of both time *and* space of *all parties* involved in the project. As stated in Chapter 5, planning involves *what* is to be done, and scheduling is *when* it is to be done. In creating a schedule for the project, the general contractor must not only manage (i.e., schedule and coordinate) its own work, but also schedule and ensure the timely execution of the responsibilities of *all* parties to the contract (see Figure 6.1). Thus, scheduling cannot be a drafting function carried out in a vacuum. It can be the organized energy behind a successful project, or the doldrums behind a failed project.

Scheduling involves the management of the general contractor's and others' *time* and *use of space*. It also involves managing activities and decisions which may influence the effective use of time and space.

By "management" (in the project sense), we mean:
- Planning *what* is to be done
- Planning *who* is to do it
- Planning *how* it is to be done
- Planning the *standards* of acceptance
- Scheduling *when* it is to be done
- Scheduling the *sequence* by which it is to be done
- Causing *decisions* to be made in a *timely* way
- Causing *project delivery* to be *timely*
- Causing the work to be *installed* in a *timely* manner
- Causing the work to be *done right*
- Causing *Contract Management* and *Change Management* to be done properly

- *Preventing problems*
- *Investigating and resolving problems* when they occur
- Giving the owner what the *contract calls for*
- Making a *reasonable profit*

The above is a mouthful—and the list is at best only partial. Successful planning will identify *all* of the steps in the process of project management scheduling. Like a road map, it identifies the route to be taken, the possible road blocks and detours, and interim destinations along the way. Successful scheduling lets all the players or team members know when they are to perform, how long they have to perform, and the constraints to performance. Successful scheduling goes one step further to manage (cause the attainment of) everyone's time line (or lifeline) of performance.

This chapter, then, is not about the specific techniques of scheduling. Much has been written elsewhere about CPM and other scheduling techniques. (The reader is referred to *Means Scheduling Manual*, by F. William Horsley for a step-by-step illustrated handbook for scheduling, and to *Project Planning and Control for Construction* by David R. Pierce, Jr., for a comprehensive guide to managing the mechanisms of the scheduling process. Both published by R.S. Means.)

This chapter on scheduling is about the *management* of:
- The players and their responsibilities
- Time
- Logistics

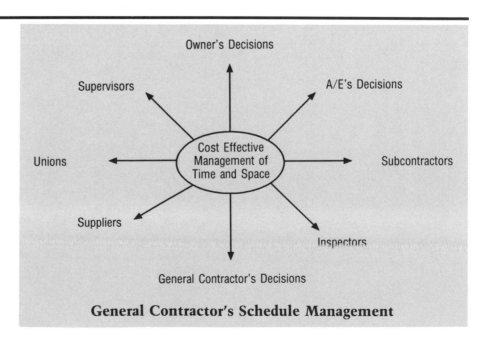

General Contractor's Schedule Management

Figure 6.1

Step 1: Read the Contract

In most contracts, the owner sets forth the details of how the job is to be scheduled. A Critical Path Method schedule may be specified, and if so, the number of activities and other requirements will be set forth. *Read these requirements and comply with them!* Know what scheduling duties are yours. Also, become familiar with:

A. *Float and who owns it.* If the contract is silent on this point, submit your schedule with a legend that reads: "The contractor owns float."

B. *Claims relating to owner's duties,* such as response to requests for information, providing owner-furnished equipment, etc.

C. *Claims relating to liability for delays,* such as:
    • Suspension of work
    • Changes
    • No Damages for Delay
    • Terminations
    • Liquidated damages

D. *Clauses that entitle the contractor to extra compensation for delays,* and the required notification and documentation.

Step 2: Scheduling the Responsibilities of Others

Once the general contractor has read the contract and becomes familiar with his and the owner's duties, he needs to take these duties into consideration in both the initial and ongoing scheduling process. The checklist in Figure 6.2 shows the duties of owner, general contractor and subcontractors regarding scheduling. (Remember, these duties may vary from contract to contract, so *read the contract* and develop your own scheduling checklist for each project.)

Step 3: Information Transfer from Estimating

If you have not already done so, obtain relevant scheduling information from the estimators and discuss with the head estimator how he priced:
    • General conditions & project administration
    • Early or full duration completion
    • Adverse technical conditions, such as dewatering
    • Weather
    • Winter work
    • Material & equipment delivery problems
    • Production flow
    • Other factors that have some bearing on scheduling

Step 4: Contract Duration

The contract generally provides for a schedule (duration), and possibly liquidated damages if the contractor fails to meet that schedule, unless delayed for *force majeure* (Acts of God) reasons. So, the important step in scheduling is to make sure the contract schedule commitment can be met.

NTP---------------DURATION---------CONTRACT COMPLETION

(Notice To Proceed)

Of course, you begin this process by RTC (reading the contract) and making sure you have all the bases covered.

Step 5: Early Completion

The contractor may have determined during the estimating phase that the project could be completed early. If the contractor can demonstrate from its bid file that it developed its pricing based on an early duration schedule, and if it schedules its work (and the subcontractors' work) on the basis of an early duration schedule, and if the contractor is delayed beyond its early duration schedule (which was communicated to the owner), the contractor may have a claim for delay damages for the period after the contractor's early duration completion work. So, the next step in scheduling is to determine if an early completion date is warranted.

Scheduling the Duties of Others	
Item	**How much time does the owner have?**
1. Submittal response	1. Per the contractor's submittal schedule. Unless the contractor is at fault for a late submittal, the owner must never take so much time that late delivery of equipment affects the installation process.
2. RFI (Request for Information) Response	2. For legitimate Requests for Information/Clarification, the owner must respond in sufficient time to prevent an impact to the installation process. Contractor must inform owner (architect) of that date.
3. Owner-furnished equipment/drawings	3. Per the contractor's schedule, but never beyond a date that will adversely affect the installation process.
4. Other approvals (e.g., blasting plans, dewatering plans, etc.)	4. In sufficient time for an adequate review, so long as it does not impact the contractor's field operation.

Figure 6.2

Step 6:
Procurement
Schedule

After the duration has been established, the next step is to schedule the delivery of material and installed equipment. This is the Achilles heel of most contractors, for scheduling the logistics is complex and involves many parties. What are the elements to be considered in scheduling the logistics? Figure 6.3 shows the procurement process for an equipment item.

The construction industry and all its members must understand that most manufacturers will not begin the fabrication process *until the shop drawings have been approved and received by the manufacturer.* Consequently, changes to shop drawings, or comments by the architect which require re-submittals can delay the start of fabrication. The manufacturer may very well place another order ahead of yours—and delays can become exponential. For this reason, the shop drawing approved date must not only be scheduled, it must also be *managed.* See Figure 6.4 for a diagram of the submittal process, and Figure 6.5 for a checklist of activities involved in scheduling the submittal process.

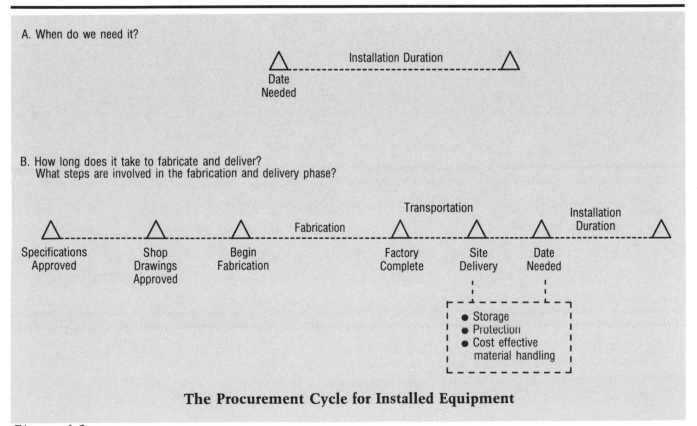

The Procurement Cycle for Installed Equipment

Figure 6.3

Step 7: Production Flow

Each trade has a "production flow," that is, a sequence of how work crews move through a building. A typical construction process is comprised of both vertical and horizontal relationships. To illustrate, a drywall operation on a high-rise building project is shown in Figure 6.6.

The horizontal axis gives two time references, the beginning and the end of the project, and the duration of each task (i.e., layout, framing, etc.). The vertical axis shows the *interrelationships* between each work area. For example, it shows that at any given time, there are only two layout crews. Once this information has been obtained, the pricing of manpower and material handling is greatly simplified. However, it should not be forgotten that this, too, is an essential step in scheduling.

When we have established how the job is to be built, we have determined the *production flow*. Production flow is important for many reasons:

- It develops the most efficient use of the work force and material handling practices.
- It shows the project manager where a crew should be and where it should go. It also provides other craft contractors and the owner (and the architect) with a visual reference so they will know when their actions may be interfering with another's "production flow."
- When the production flow is followed, an *effective learning curve* can be developed.
- It is a useful tool for demonstrating *impact* in a claims situation.

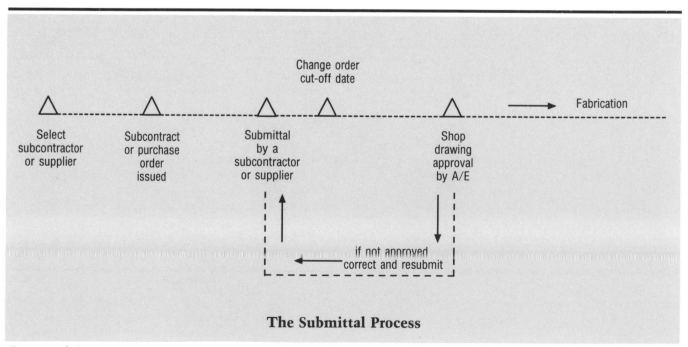

The Submittal Process

Figure 6.4

Checklist for Scheduling the Submittal Process

Scheduling the submittal process requires the following activities:

☐ Contracting with the subcontractor or supplier to meet both submittal and delivery dates.

☐ Requiring the different suppliers of equipment on complex projects to *coordinate* their submittals prior to submittal to avoid installation problems or conflicts.

☐ Submitting a construction schedule to the owner or architect which demonstrates clearly the turnaround time on each submittal.

☐ Submitting a construction schedule to the owner or architect which shows when all changes to the equipment must be completed (cut-off date), so that the delivery and installation schedules will not be jeopardized.

☐ Establishing *critical priorities* for submittal turnaround with all members of the project. For example, it may be that on the front end of a project, rebar must be turned around in three days, whereas electrical fixtures may tolerate a sixty-day turnaround time.

☐ Never allowing the mail to slow or stop the process. Get on the phone and work out any comments.

☐ Letting others know the consequences of the delay—both in time and money.

☐ *Staying on top of the submittal process!* In the submittal process, the general contractor must:

Schedule:

_____ the submittal process

_____ equipment deliveries

_____ installation duration

Manage:

_____ subcontractors

_____ suppliers—including backlog and coordination with other suppliers

_____ architect and engineer—includes comments, changes, and delays

☐ Make it happen!

Figure 6.5

Step 8:
Relationship with Other Crews

In step 5, we demonstrated the production flow of one trade contractor. In step 6, we show the relationship of contractor to contractor. As previously recommended, these schedules are the result of meetings with and input from all the key players on the team.

Again, the general contractor must not only schedule in this manner, but he must *manage* the schedule so that all the team players are following it and not interfering with its smooth flow. The general contractor must make a reasonable effort to create and maintain this production flow. This is the essence of management and the key reason an owner hires a general contractor. As the project manager is building the schedule, he must figure out how to get the other team members to "buy into" this plan.

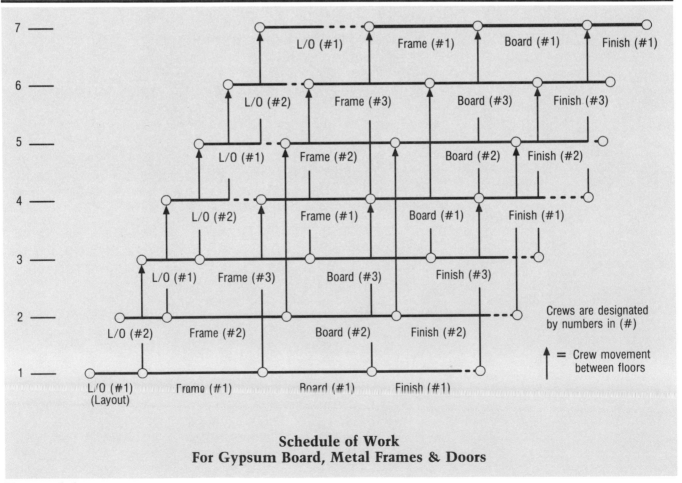

**Schedule of Work
For Gypsum Board, Metal Frames & Doors**

Figure 6.6

Step 9: Float—Whose Is It?

Owners and architects often believe that they own the project's float. However, unless the contract says otherwise, owners/architects do not own the float. The contractor probably does not own it either, unless he has negotiated this point through a clause in the contract. Most authorities maintain that float is "shared." We suggest that as a part of the scheduling activities, the general contractor try to come to grips with the amount of time and space the owner and architect may have that they can use without unreasonably delaying the project (taking into consideration the possibility of *force majeure* conditions).

In other words, for all of the above, the owner/architect has a *"window,"* or *time frame* for decision making. *The contractor must schedule these windows or time frames*, and then communicate those limits to the owner/architect. He must then manage the owner to ensure performance of these functions within the scheduled time frames. The use of time, or float, through changes will be discussed in more detail in the next chapter.

Step 10: Schedule Your Own Resources

The contractor must not only schedule the activities of others, but also plan the resources for which it is responsible. This includes assigning the:

- superintendent
- foremen
- work force
- rolling stock equipment
- cranes
- small tools
- trailer

The Short Duration Schedule and "Look Ahead" Schedules

In addition, on jobs with a duration of a year or more, we recommend developing a short duration schedule (such as a ninety-day schedule) and then (working with subs and suppliers and the owner) to finalize the complete schedule. Furthermore, two- or three-week "look ahead" schedules should be developed for use at coordination meetings.

How to Use the Schedule

The fundamental steps in the management of scheduling have been presented in the previous sections. The following discussion deals with the issues that arise once a schedule already exists and the project is under way. How does the general contractor schedule and control the project in the field? The following are guidelines for this phase of scheduling.

Effective Coordination Meetings with Owner and Architect

- Meetings should be held routinely and as required to identify priorities that must be acted upon and decisions that must be made.
- Attendees should include: the owner or his designated representative, the lead architect and key consultants (such as electrical or mechanical), and the contractor's project manager

and superintendent. It is a good idea to have the construction firm's president or vice-president attend these meetings occasionally.
- A good set of minutes should be maintained, and should include any action items and need dates.

Effective Coordination Meetings with Subcontractors and Contractor's Own Supervisory Personnel

- Hold weekly meetings; keep a good set of minutes.
- Develop "look ahead" schedules
- Ask and act upon the right questions as to what space needs to be ready for follow-on work
- Review changes and their effect on the work
- Review submittals of material and equipment
- Review work force
- Review material handling procedures and restraints
- Review restraints to job progress and how to improve
- Review quality requirements and workmanship problems
- Review safety requirements.

Decision-Making

Decision-making is a major factor in effective scheduling. Effective decision-making by the construction firm's own management, as well as by the owner and architect/engineer, must be scheduled, and should be taken into account as an activity when constructing the schedule.

The decision-making process may be the most ill-managed, least understood system in the construction process, and the result is a tremendous number of avoidable problems. For example, most equipment delivery problems are directly associated with decisions regarding differing site conditions, changes, or requests for clarification.

To a considerable extent, construction cost escalations and schedule delays are a result of the decision process, or lack of it. What is meant by an effective decision process?—making a decision (based on an objective review of all available facts) in a timely way so as not to exceed the project's cost or schedule. It should be noted that decision-making is not just the final act of the decision, but the *process* that leads to the necessity for a decision.

Making decisions is a mutual process, involving the decision-maker and the party who requires the decision. The party needing the decision must provide information that is adequate and reliable enough for a decision to be made effectively and objectively. The decision-making process in each instance requires an active role by both the decision-maker and the party requesting the decision and providing the information.

Failure to Decide

The various parties to a contract sometimes become so embroiled in assigning the blame for a problem that they lose sight of the fact that there is a problem to be solved and a project to be built, and time is slipping away. In a case involving an interchange of construction, the owner and contractor argued for months about who had the responsibility for adding approximately $50,000 worth of rebar to the pilings to prevent damage as they were being driven.

Because of the delay, the owner wound up compensating the contractor for over $3,000,000 for the resultant impact to his operations.

How to Ensure Timely Decision-Making

The general contractor is typically hired to schedule and coordinate the work. Inherent in these duties is the responsibility to oversee the decision-making process of all the parties involved in the project. The proactive general contractor must find a way to ensure that *all* parties perform their decision-making duties. It is important to realize that if anyone in the decision-making process fails, the system fails. Figure 6.7 is an example of the chain of events in a simple, but common situation. It is clear that if any one of the links is broken, if anyone fails to follow through, the decision-making process comes to a halt, and time slips away. It is up to the general contractor to ensure that all involved do their part.

How does the general contractor carry out this task? First, he must inform others of their duties and what is expected of them. To accomplish this, he may take the following steps.

- At the outset of the project, meet with the architect and owner and present them with priorities and procedures regarding turnaround of submittals, requests for information, and changes.
- Have a similar meeting with major subcontractors/suppliers, again specifying requirements for their handling of paperwork. Tie these requirements into the subcontracts and purchase agreements.
- In the construction contract and schedule, include the decision-making duties (e.g., submittals) of all parties, with specific dates and/or guidelines.
- Show change order cut-off dates by craft on the schedule.

Second, the general contractor must accept responsibility for decision-making as an aggressive, proactive function. He must stop saying: "The architect is not providing responses to submittals on time," and instead say: "I am not managing the architect in such manner as to cause him to issue decisions on a timely basis."

Third, the general contractor must let others know in writing the consequences of their failure to perform the decision-making function on time. This should be done in a clear and specific way, as in the following excerpts from sample letters.

- "The failure to approve the pre-cast load calculations by July 1st will cause a late delivery of the pre-cast because of the supplier's backlog. If this occurs, we will be thrown into a period of potentially inclement weather which may increase both the time and the cost of performance."
- "We must have an architect's representative on site due to conflicts in the drawings. Our company bid on one mobilization and re-mobilization for each activity. Because of the delays in getting responses, we are having to pull off our crews and relocate them elsewhere, then come back when we get an answer. As a result, we have, in some cases, several more mobilizations and re-mobilizations than we bid or had reason to expect."

General contractor is informed that countertop/lavoratory units will not be available for delivery for six more weeks due to a strike. Needs direction on how to proceed.

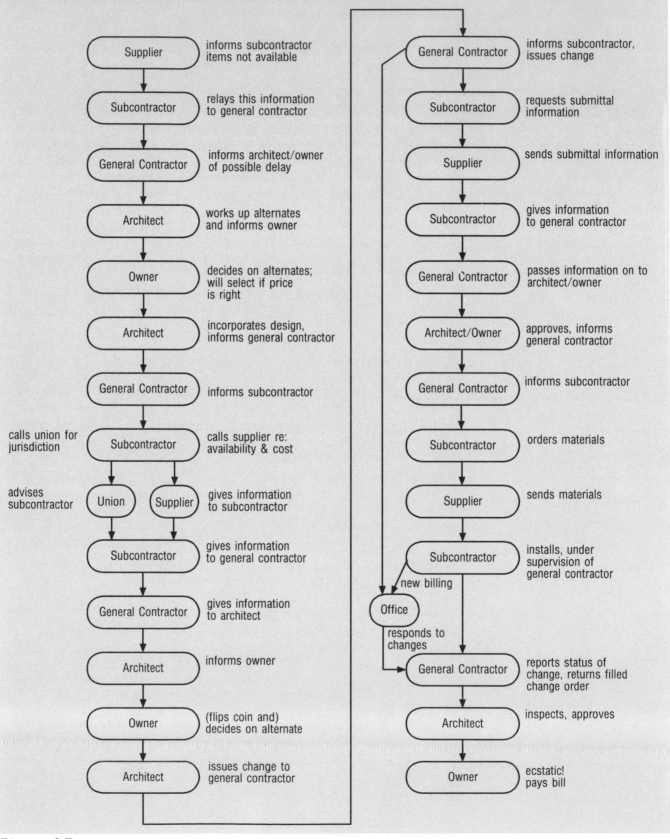

Figure 6.7

Fourth, the general contractor must provide adequate information for others to make a decision. The following items are typical of the types of information the general contractor should provide to others for certain situations.

For Differing Site Condition:

1. The contract states that the owner will provide direction when a differing site condition is encountered.
2. We must have that directive by *(date)*, or the foundation will be delayed.
3. The log borings contained in the contract documents show water to ten (10) feet above the planned inverted elevation.
4. The soils reports indicate static water, which is pump-controllable.
5. Our estimate (Attachment No. 1) was on pump controllability.
6. In fact, there is water under pressure (or artesian water) which cannot be controlled by pumps, and which may cause subsidence to the site.
7. A report by Water Consultants, Inc. substantiating our position is attached.
8. We appreciate an immediate response providing direction.

For Disputed Charge:

1. Specification # ____ provides a minimum dowel length of 22".
2. Drawing # ____ shows the dowel length to be 22".
3. The contract is read as a whole.
4. The contract does not impose any design duties on the contractor.
5. We submitted shop drawing showing dowel length at 22". This submittal was approved for construction by your engineer [See attached __(No. ___)__].
6. Attachment # ____ contains statements from other bidders that they bid on the basis of a 22" dowel length.
7. The pilings are failing. The consultant's report shows our quality control to be satisfactory. It also shows that we are complying with the contract, and the consultant's opinion that the problem is the inadequate length of the dowel.
8. Activity No. ___ has been delayed while awaiting your decision. All other successive items on the critical path will also be delayed. Our daily time sensitive expense is $ ____, standby equipment is $ ____ daily, and loss of productivity could be substantial.
9. We request a meeting immediately to obtain direction as to how to resolve this problem.

In both these examples, the contractor is providing the owner with the following:

- His contractual duty in this issue.
- The facts of the situation.
- The factual basis of the change from the contract.
- The impact of the current situation.
- The impact of the failure to decide.

He is also asking for direction in the form of a decision. Whether the contractor likes it or not, most contracts provide a vehicle for decision-making, detailing who is responsible and how much is to be decided later. Therefore, the contractor should:

- Push for directions, which the owner has the responsibility to provide.
- Document the steps taken and the additional costs.

Fifth, the contractor must practice what is called "Finger-in-the Chest" management. In other words, the contractor must:

- Inform the party of its duty to decide.
- Tell the party *when* the duty should be performed, i.e., when the decision must be made.
- Remind the deciding party at least once of the upcoming date (e.g., a letter notifying the architect of outstanding submittals a week before due).
- As soon as the date has come and gone for decision-making, inform the party responsible for deciding in writing.
- Ask for a meeting with someone in authority if a delay lasts very long. In other words, move up from the representative with whom he is currently dealing to someone who truly has the authority to decide.

Sixth, the general contractor must keep these **Don'ts** in mind.

- Don't be emotional.
- Don't name-call.
- Don't rely completely on written correspondence. Set up a meeting or conference call to address the problem right away.
- Don't let the project fall behind while blaming others for failing to make a decision.
- Take all possible action in order to stimulate a decision. In some cases, it may be necessary to tell the owner what you intend to do unless he objects, and that you intend to get paid for it.
- Do not allow the architect to stop a project by adding notes on submittals (that can be incorporated into the manufacturer's final engineering) without reprocessing the submittals. Upholding this principle can be difficult, and it helps to educate the architect at the beginning as to the potential impact on a project when he fails to "AWC" (Accept Without Changes) submittals, and that by disapproving and causing resubmittal, he holds up the procurement process. Specifically, problems arise when the architect:
  - Fails to provide written notification to the decider.
  - Fails to document the contractor's costs.
  - Fails to update the contractor's schedule.
  - Fails to allow for a time extension.

Clearly, the project schedule can be greatly affected by poor decision response time. By proactively managing the decision-making process, the contractor can ensure that decisions are made in a timely fashion and, therefore, that decision-making does not have a negative impact on the schedule. Proactive management can help the contractor "brace its fence" against risks to the schedule due to indecision.

The Schedule and the Fence

The contractor's job is to protect the fence, or bottom line project cost. It must do this by protecting:

- The duration of the project
- The sequence of work
- The production flow

Now, how can encroachments into the schedule fence be prevented, or their effects be mitigated? First, there must be a firm commitment to updating the schedule on a regular basis. Second, cut-off dates should be established for change orders and enforced; it should be agreed that change orders occurring after these dates will impact the schedule. Third, subcontractors must be coordinated carefully.

Commitment to Updating the Schedule

The critical path scheduling method and computers are wonderful tools that can be used to perform updating in a minimum time. Contractors who contend that numerous changes have prevented them from accomplishing updates have a poor excuse. As soon as schedule updates begin to be neglected, the contractor loses track of where the job is, where it should be, and where it is going; at this point, the contractor has lost control of the job! When this occurs, costs can increase at more than an arithmetical rate. This is why updating the schedule is so crucial.

Contractors should maintain detailed cost records showing unit cost. They should also develop "unit schedule" standards which can provide an index, or rule-of-thumb guideline for change order impact. The development of rule of thumb guidelines for nonrecurring tasks (e.g., setting up equipment for pile driving) and recurring tasks (e.g., the actual pile driving) is recommended. These standards can be used effectively in setting up the original schedule, determining schedule impact, or updating schedules in response to change orders or for comparative purposes in claims situations. They also serve as excellent pricing tools.

Subcontractor Management

Coordination is the key to subcontractor scheduling. In order to coordinate, the contractor must:

- Be familiar with the activity and schedule of the preceding subcontractor.
- Make sure the preceding subcontractor knows the work and the schedule for which he is responsible.
- Make sure he carries it out accordingly.

Summary

All of the parties to the project have duties regarding scheduling. In order to ensure that decisions are properly made by these individual parties, the contractor must have an overview of each of their areas of responsibility. Figures 6.8a and 6.8b summarize the schedule-related duties of the general contractor and subcontractors.

General Contractor's Schedule-Related Duties

☐ To pre-plan at the beginning of a job.

☐ To comply with contractual schedule requirements.

☐ To make a complete schedule, including:

    ____ all activities

    ____ procurement schedule

    ____ deliveries/lead time

    ____ interrelationship of activities

    ____ input from field

    ____ input from subcontractors

☐ To update the schedule as required.

☐ To plan work weekly and daily.

☐ To maintain adequate, competent management and labor on the job.

☐ To coordinate subcontractors.

    ____ keep preceding subcontractors "out front."

    ____ have area prepared for following subcontractors.

☐ To cause work to be performed.

☐ To be pro-active in decision-making.

☐ To define owner's timely duties.

☐ To mitigate damages to owner and subcontractors.

☐ To have a quality control plan to avoid re-work.

☐ To get submittals in and approved on time.

☐ To have a comprehensive material list and to provide material on a timely basis.

☐ To provide adequate material storage and handling.

☐ To check materials for quality and specifications as it arrives.

☐ To check for dimensions/congestion prior to installation.

☐ To plan close-out.

☐ To plan punch lists.

☐ Turn-overs.

Figure 6.8a

Subcontractor's Schedule-Related Duties

☐ To provide required input to general contractor regarding schedule.

☐ To know job conditions:

___ to inform general contractor of significant job conditions affecting scheduling.

___ to communicate its plan of action to other crafts and find out what their plan is.

___ to acquire delivery/lead times.

☐ To plan its own work.

☐ To stay ahead of other crafts.

☐ To supply on time:

___ submittals

___ equipment

___ materials

☐ To maintain adequate and competent supervision and manpower.

☐ To perform quality work to avoid rework delays.

Figure 6.8b

Chapter Seven

Management of Subcontractors

Chapter Seven

Management of Subcontractors

Managing subcontractors is one of the primary functions of a general contractor. The quality of that management is a major determinant of the general contractor's success. This chapter presents the basic aims of successful subcontractor management, and backs them up with checklists that ensure a proper approach to all phases of the general contractor's dealings with subcontractors.

Owner/General Relationship

The relationship between the owner and the general contractor is one of privity of contract. That is, it is a one-on-one contractual relationship, in which the general contractor has a duty to the owner to perform the work set out in the contract documents, within the specified time and quality standards. The owner looks to only one party—the general contractor—*who has complete responsibility for contract execution and compliance.*

The general contractor may "sub out" some of its work without altering its duties or responsibilities to the owner. The general contractor is still completely obligated to the owner for total contract compliance (see Figure 7.1). By subcontracting, the general contractor has, however, created a new set of legal obligations, as shown in Figure 7.2.

General/ Subcontractor Relationship

Establishing and maintaining a firm, but cooperative relationship with subcontractors should be a policy of the general contractor. Encouraging such cooperation enables all parties to make money on the project, and can lead to other mutual benefits as well. For example, subcontractors may work with the same general contractor on future projects, and working effectively together as team members may help the general contractor to be competitive as he provides a more effective service to the owner. There seems to be a major trend toward this kind of team building and the reduction of adversarial relationships on construction projects.

Fair Treatment

At the outset, it is important to note that general contractors have a duty to treat subcontractors fairly. The word *fairly* may seem subjective and therefore difficult to define, but it is helpful in categorizing the obligations of the general contractor, as shown in Figure 7.3.

If the general contractor treats the subcontractor unfairly, the subcontractor may possibly be successful in suing the general contractor for damages, including punitives (or treble damages under some statutes). The enforceable nature of the duty to treat the subcontractor fairly is a growing trend. It is up to the general contractor to seek the best possible professional counsel regarding the legal ramifications of this issue in the local area where a project or projects are to be constructed.

How to Manage Subcontractors

The subcontractor is an *independent contractor*, not an *employee* of the general contractor. The general contractor contracts for the subcontractor's management skill and expertise, just as the owner contracts for the general's scheduling and managerial expertise. The general, then, must direct the subcontractor's job of managing, rather than directly overseeing the subcontractor's labor forces. This is similar to the way in which an owner directs the general contractor's management efforts. How *does* the owner manage the general contractor's management (at least, how do the good ones do it)?

Owners generally operate on the basis of **expectation**. That is, owners let general contractors know what they expect of them. To begin with, owners expect contractors to be experienced and qualified. Then, owners set forth, as clearly as possible, what they expect in terms of the *scope of work* to be done, its *standards of acceptance, when the work is to be performed,* and *who has the financial risk* associated with its performance. The clearer these expectations are, the greater the probability for job success.

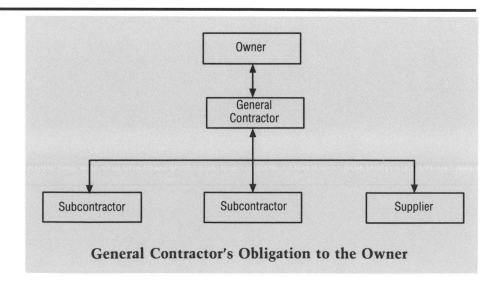

General Contractor's Obligation to the Owner

Figure 7.1

In the next phase, the owner enforces his expectations. This is done by overseeing the contractor's work and forcing him to do it correctly. The owner has some very effective tools at his disposal that can be used to his expectations. For example, the owner can withhold the contractor's payments or even terminate his agreement with the contractor if the work is not correctly done. In other words, the owner establishes *expectations*, then *observes* the contractor to see that those expectations are fulfilled, and *enforces* those expectations through the exercise of payment and termination clauses. The owner then *rewards* (pays) the contractor for meeting those expectations. Figure 7.4 is a checklist summarizing the various facets of the owner's management of the general contractor.

Where Do General Contractors Go Wrong?

Historically, general contractors seem to fail in the management of subcontractors in the key areas listed below and described in Figure 7.5.

- Bid Phase
- Contracting Phase
- Scheduling
- Contract Compliance
- Communications
- Coordination

Contractor's Legal Obligations Resulting from Subcontracts

General contractor's contractual duties to the owner...
- remain unchanged as a result of adding a subcontractor;
- include responsibility for the subcontractor's unexcused non-performance.

General contractor's contractual duties to the subcontractor include...
- scheduling the work;
- coordinating craft contractors' work;
- causing precedent trades to perform their work on time and per the contract;
- providing decisions on a timely basis;
- treating subcontractors fairly;
- paying subcontractors on time for work complete or materials on site as per the subcontract agreement.

General contractor's duties in tort (negligence) to third persons may include responsibility to:
- subcontractor's workers who may be injured due to unsafe job conditions;
- passersby (or other third parties) injured by subcontractor's negligently performed work or operation.

Figure 7.2

Fair Treatment of Subcontractors: Checklist of Key Elements

The general contractor's duty to treat subcontractors fairly includes:

☐ Disclosing all pertinent pre-bid information to the subcontractor.

☐ Distributing all requests for estimates to all subcontractors to review for impact—in a timely manner.

☐ Permitting the subcontractor to participate in change order negotiations, not denying him his "day in court."

☐ Keeping subcontractors up to date on schedule changes.

☐ Representing the subcontractor objectively and fairly to the owner.

☐ Paying the subcontractors in a timely manner for work completed as per the contract agreement.

Figure 7.3

How an Owner Manages a General Contractor

An owner manages the general contractor's management by:

☐ Selecting contractors who are experienced and qualified, both technically and financially.

☐ Defining the job to be done with detail and clarity.

☐ Scheduling for performance:

    _____ submittals

    _____ interim schedule

    _____ financial completion

☐ Establishing control mechanisms for:

    _____ shop drawing approvals

    _____ progress approval

☐ Establishing requirements for:

    _____ quality control program

    _____ safety program

    _____ written notifications

☐ Kicking the job off with a pre-construction conference.

☐ Routinely inspecting work against consistent standards.

☐ Directing contract compliance.

☐ Paying in a timely way for work actually installed.

☐ Answering *all* project queries completely and in a timely fashion.

☐ Attempting to create a team.

Figure 7.4

Management Failures to Avoid Checklist

Bid Process

☐ Failure to work closely with subcontractors to inform them of bid terms and conditions, etc.

☐ Use of technically weak or inexperienced subcontractors.

☐ Failure to check references to avoid use of financially weak, unbondable, or unreliable subcontractors.

☐ Failure to realize full scope of subcontractors' work

Contracting Process

☐ Failure to write a scope of work that fully covers the job to be done by the subcontractor and work to be done by the general contractor.

☐ Failure to include adequate information regarding submittal data.

☐ Failure to include delivery dates.

☐ Confusion as to interface between trades (e.g., who hooks up the mechanical equipment—the mechanical or electrical subcontractor?)

Scheduling

☐ Failure to include subcontractor's input in original or updated schedule.

☐ Not requiring the subcontractor to plan his own work and to submit resource plans on a weekly basis.

☐ Not requiring the subcontractor to plan close-out of its own work.

Contract Compliance

☐ Not checking out the subcontractor's quality of work as it progresses.

☐ Not checking submittals before they are submitted to the owner's representative.

Communications

☐ Failing to keep subcontractor informed of all changes.

☐ Failing to answer questions in a timely way.

☐ Not listening to subcontractors.

☐ Failing to involve the subcontractor in meetings and/or job discussions.

Coordinating the Project

☐ Not allowing the subcontractor any input as to plans of action.

☐ Not allowing the subcontractor input into methods of previous contractor.

☐ Failure to keep the subcontractor on the schedule.

Figure 7.5

Bid Phase

The general contractor's coordination and management duties during the bid phase are as important as they are in the construction phase. Too often, the general makes the mistake of not getting involved with the subcontractors during the bid activity. Many disputes and problems that occur during construction could have been (and should have been) prevented during the bid phase. Figure 7.6 is a checklist for the general contractor to use in managing subcontractors during the bid phase. The Subcontract Bid Review Checklist in Figure 7.7 will aid the general contractor in choosing the right subcontractors for the job.

Keep in mind that fax machines, once considered a luxury, are appearing in more and more offices, where they have become as common as copy machines. A fax machine is a helpful tool at the location where bids are being taken. Bids received by telephone can be confirmed instantly without the "I'll drop a copy in the mail" delay. A fax machine can also be very handy if a clarification is needed on a detail—either pre-bid or during construction. Details can be quickly fax'd, saving the time it would take to mail or send sketches by messenger.

Checklist for Subcontractor Management During the Bid Phase

The checklist in Figure 7.6 is designed as a tool to help the general contractor manage subcontractors during the bid phase of a project. The operative word is *manage*. During the bid phase, general contractors too often simply take in and compile telephone or written quotes from subcontractors and suppliers. Sometimes general contractors base their estimate for a job on the low bid of a critical subcontractor whom they have never met or checked out. *The low bid is everything!* Often there may be important language in the general conditions (such as pre-bid approval of "or equal" equipment) which is not included in the technical specifications, yet the general contractor does not make sure that the subcontractor is aware of these general conditions requirements.

The point is, of course, that the entire subcontractor bid process must be *managed*. The bid phase must not be simply an exercise in quote-taking by the company receptionist!

The checklist in Figure 7.6 highlights key risk areas in the bid/estimate phase which must be managed. The head of the estimating team has the responsibility for determining the qualifications of the potential low bidder (among subcontractors), and must also ensure that the bid is responsive and complete.

Subcontract Bid Review Checklist

The Subcontract Bid Review Checklist (Figure 7.7) is a tool that the general contractor uses to evaluate fully the subcontractor's bid. This enables the general contractor to select the *correct*—not just the low—contractor bid. It is better to know the problems and risks of the subcontractor before you sign the contract than to discover it during a crisis that the subcontractor created on the project. Both the estimator and the project manager should be involved in this process of subcontractor evaluation. Remember, taking bids is not just a clerical function. Furthermore, simply

Checklist for Subcontractor Management During the Bid Phase

During the bid phase, the general contractor must make sure that subcontractors:

☐ ... are aware of relevant general contract provisions, i.e., milestone dates, penalty clauses, warranties, etc.

☐ ... are bidding on *all* work that the general contractor expects the subcontractor to perform, regardless of where it is recorded in the contract documents.

☐ ... have the appropriate financial capability, experience, and integrity before including their quote in the bid.

☐ ... can provide a bid bond in cases where the value of the subcontract gives the general contractor an unacceptable level of risk.

☐ ... see all pertinent plans, specs, and addenda.

Figure 7.6

Subcontract Bid Review Checklist (Low Bid vs. Correct Bid)

	Yes	No
1. Has this subcontractor had previous quality problems?	____	____
2. Is this subcontractor slow to perform work?	____	____
3. Does this subcontractor resolve problems or litigate?	____	____
4. Is this subcontractor being selected on price alone?	____	____
5. Is this subcontractor financially sound?	____	____
6. If not, what steps are being taken to protect our interests?	____	____
a. Reject the subcontractor?	____	____
b. Bond?	____	____
c. Joint checks (in both sub's & suppliers' names)?	____	____
d. Increase retainage?	____	____
e. Are these steps being written into the agreement?	____	____
7. Does this subcontractor have a union security agreement, if required?	____	____
8. Are there conditions on subcontractor's bid which must be resolved?	____	____
9. Is subcontractor's bid complete?	____	____
10. Is subcontractor's bid responsive?	____	____
11. Can subcontractor meet schedule?	____	____
12. Can subcontractor effectively work with CPM schedule?	____	____
13. Is this a critical path subcontractor?	____	____
14. Is it clear which services (crane, hoist, storage, etc.) the general will furnish to the subcontractor?	____	____
15. Does the subcontractor have performance or design responsibilities?	____	____
16. Is the subcontractor's surety sound?	____	____

References:

Firm	Date Called	Spoke To

1. _____

Comments: _____

_____

2. _____

Comments: _____

_____

Figure 7.7

accepting the low bid is a form of gambling. Analyzing both the subcontractor's bid and his capability to perform is *risk management.*

Another point to remember is that just because there may be some risks associated with the low-bid subcontractor does not in and of itself mean that that subcontractor's bid should be rejected. It does mean that the general contractor should be aware of the problems and figure out in advance how to manage them.

For example, it may be worthwhile to provide the critical path subcontractors with a short orientation course on how to work with a CPM. This should be done at the outset of the project, and should include a session on how craft contractors should work and communicate with each other. Again, management of these risks means confronting and doing something about them in advance.

Using the Bid Re-cap Sheet in Figure 7.8 as a checklist can help minimize risks during bidding, to ensure that items do not fall through the cracks, and to make sure that all subcontractors are bidding on the same items. To create this checklist, the following steps must be taken.

Step 1: Read the plans.
Read the specifications.

Step 2: Make a list of all items, particularly specialty items or unusual details that the subcontractors must include in their bids.

Step 3: Use or make a bid summary sheet. List all items that must be included in the bids down the side of the sheet. Also include a space for alternates or cost cutting suggestions.

All expected bidders, as well as a contact person and telephone number, should be listed across the top of the page. (Figure 7.8 is an example of a completed bid checklist.)

Step 4: As bids come in, use this sheet to perform a quick review with the subcontractor. This review will ensure that all subcontractors are bidding on the same items. If they have the item in their bid, simply check it off. If not, circle the item. You will have to add a cost for this item during the bid review process. The cost for the missing items can be obtained from the prospective subcontractor, other prospective subcontractors who can give you an idea of the value of the item, or your own estimate.

This process will help you decide whether the apparent low bidder is the actual low bidder.

In the sample checklist in Figures 7.8a & b, the apparent low bidder, as it turned out, left out some essential site work items. When compared on an equal basis, it was discovered that the apparent low bidder was not the actual low bidder. The omitted items would have to be included somewhere and if not caught now, would probably be covered under a future subcontractor change order, and paid for at a higher rate.

Means® Forms

CONDENSED
Bid SUMMARY

SHEET NO. 1/1

PROJECT Gibson Office Bldg. #321

ESTIMATE NO. .615-90

LOCATION

TOTAL AREA/VOLUME

DATE 1/27/90

ARCHITECT One Line Design

COST PER S.F./C.F.

NO. OF STORIES

Bid Section: 03-Concrete EXTENSIONS BY:

CHECKED BY:

		ABC Conc	Krackup & Brache	Miracle Works Inc	Bildenhope Co.	Budget
	Contact	K. Kong	Pete Sakes	Jack nJille	Juan Motime	
	Phone #	555-1928	555-2189	555-3214	555-4912	
	Base Bid					
	Does bid include: Conc Forms Reinf.					
	A Bolts - Material Labor					
	Finish Cure Protect					
	Special Seal @ Lab					
	Joints					
	Driveway approach aprons					
	Topping at Mezzanine					
	Precast floor/deck					
	Flagpole Bases					
	Sidewalks					
	Ramps					
	Curbs					
	Stairs/Fill pans					
	All taxes, fees					
	Per plans/specs					
	permits					

Figure 7.8a

⚓ Means° Forms

CONDENSED ESTIMATE SUMMARY

SHEET NO. 1/1

PROJECT _Gibson Office Bldg., #321_

ESTIMATE NO.

LOCATION _San Diego_   TOTAL AREA/VOLUME   DATE _1-27-90_

ARCHITECT _One Line Design_   COST PER S.F./C.F.   NO. OF STORIES

Bid Section 03 Concrete   EXTENSIONS BY: CHECKED BY:

	ABC Concrete	Krackup & Brache	Miracle Works, Inc.	Bildenhope Co.	Budget
Contact	K. Kong	Pete Sake	Jack n Jill	Juan Motime	—
Phone #	555-1928	555-2189	555-3214	555-4912	—
Base Bid	229010	256550	244022	241663	266100
Does bid include: Conc., Forms, Reinf. ?	✓	✓	✓	✓	✓
A bolts - Material	N/A	✓ (1200)	N/A	N/A	✓ (900)
Labor	✓	✓	✓	✓	✓
Finish - Cure - Protect	✓	✓	✓	✓	
Special Seal @ lab.	✓	✓	✓	✓	✓
Joints	✓	✓	✓	✓	✓
Driveway approach Apron	0 8000	✓	✓	✓	✓
Topping at Mezzanine	0 6600	✓	✓	✓	✓
Precast Floor/Deck	N/A	N/A	N/A	N/A	N/A
Flagpole Bases	0 600	✓	✓	✓	0 600
Sidewalks	✓	✓	✓	✓	✓
Ramps	✓	✓	✓		
Curbs	✓	✓	✓		3200 in "02"
All taxes, fees	✓	✓	✓	✓	✓
Per Plans/Specs	✓	✓	✓	✓	✓
Permits	✓	✓	✓	✓	✓
Totals	249610	255350	244022		269000
Alternates	—	—	—		
$ after negotiations	246000	—	239950	237000 ✓	
Will meet schedule	yes	—	yes	yes	
			"will try"	ready to go	

Figure 7.8b

Before Awarding Subcontracts

After being awarded the contract, the general contractor should begin immediate preparations for the construction phase. Before issuing the Notice to Proceed, the general contractor should take the actions described in the Pre-Subcontract Award checklist in Figure 7.9.

The Subcontract Agreement

Experience dictates the use of the following guidelines in preparing and executing the subcontract agreement:

- *No payment* should be made to any subcontractor who has *not signed* the contract or has *not provided* proof of required insurance. In fact, good practice dictates that *no* subcontractor should be allowed on site without having submitted a valid insurance certificate. The general contractor who does not abide by this rule is courting disaster. It is unthinkable that a general, who would not dream of doing business without a detailed contract with the owner, would risk subcontracting a large, or any, portion of the work without a comprehensive signed agreement with an insured subcontractor.
- The agreement should clearly define any *performance requirements* included in the general contract that are to be met by the subcontractor.
- All subcontracts should contain *performance dates*. A subcontractor cannot be held accountable if the schedule has not been properly laid out from the beginning. The schedule should also be updated, maintained, and communicated to the subcontractor, and dates should be specified for shop drawing submittals.
- The general contractor should always have subcontractors participate in setting up the schedule, and should state this intention in the subcontract. This is a key step in promoting the powerful team concept. The schedule also benefits from the knowledgeable input of the important crafts on the project. Their participation in the schedule also gives them a sense of *responsibility* and *commitment* to meet the schedule.
- The scope of work to be performed by the subcontractor should be described explicitly in the agreement. If the subcontractor is to install a total functioning system, this requirement should be stated clearly, in specific sections of the general contract with the owner. The general contractor should pay particularly close attention to the next five guidelines.
- Where ambiguities in the general contract may be resolved in favor of the general contractor, ambiguities in the subcontract agreement may be resolved in favor of the subcontractors, since the subcontract was written by the general.
- All exceptions to the subcontractor's work should be spelled out in detail. If, for example, the general contractor is to perform excavation for the electrical subcontractor, this arrangement should be stated clearly.
- Proposals by the subcontractor should not be incorporated by reference or in full in the subcontract agreement. Differences between the proposal and the original scope of work, if any,

Pre-Subcontract Award Checklist

Before issuing the Notice to Proceed, the general contractor should:

- ☐ Notify subcontractors in writing of the intent to award them a contract in accordance with their quote.

- ☐ Conduct financial surveys on subcontractors whose capacities are in doubt.

- ☐ Decide which subcontractors will need to post performance and payment bonds, and immediately serve notice that such will be required.

- ☐ Make a file for each subcontractor in the project records.

- ☐ Place quotes, proposals, executed contracts, bonds, insurance certificates, correspondence, and other relevant data in each file.

Figure 7.9

should be resolved, and the results included in one document dictating the rights and responsibilities of the parties. Where applicable, a shop drawing schedule should also be included in the agreement.

- The statement of work should never contain extraneous and potentially problematic issues, such as *warranty*, which should be spelled out in the general conditions or in a reference to the general contract with the owner. Liquidated damages, on the other hand, should be spelled out in the agreement. Since all previous promises and representations are made null and void by the signing of the subcontract agreement, those oral warranties and promises that the general contractor wishes to retain as valid should be specifically included in the written subcontract agreement. This is where good documentation practices pay off.

- Documents that the subcontractor is required to supply, such as operating instructions, manuals, warranties, and as-built drawings, should be covered by the subcontract agreement.

- The general contractor should use the Subcontract Agreement checklist in Figure 7.10.

The Subcontract Agreement Checklist
The checklist in Figure 7.10 provides a convenient method for organizing information in order to make the decisions listed in Figures 7.6 and 7.7. It also contains standard questions that can be used to safeguard against the exclusion of necessary or advisable provisions from the agreement.

The owner, as previously discussed, establishes (or tries to) clear expectations in the contract documents regarding what he or she wants and when. These expectations involve both scope of work and administrative requirements, such as shop drawing submittals, timeliness, and the nature of written notifications. The general contractor should do no less! The following checklist to be used by the project manager is a way for the general contractor to make sure that his own key expectations are included in the subcontract.

In addition to reviewing the subcontract agreement for the items in the checklist, the following procedure is also highly recommended:

At the contract signing, have the president or other key officer of the subcontracting firm meet with the project manager (or president) of the general contracting firm to discuss the subcontract agreement and the method of operation. Time-consuming as this step may seem, it establishes a team at the outset of the project, gets the commitment of a high-ranking officer of the subcontractor, and provides a means of communication with a real decision-maker for dealing with problems later on as the project progresses.

Pre-Construction Conference

At the beginning of the construction phase, a conference with all the major subcontractors can set the tone for the job. Thereafter, weekly meetings should be held to keep track of progress, coordinate work, and point out potential problems. The checklist in Figure 7.11 lists the topics to be covered in the pre-construction subcontractor conference.

Subcontract Agreement Checklist

Does the subcontract contain: Yes   No   Date

- ☐ An adequate work statement (inclusions, "no matter where shown or specified")—that does not reference proposals? _____ _____ _____

- ☐ A list of all specifics included in the contract? _____ _____ _____

- ☐ A list of all exclusions of the work to be performed or material to be furnished? _____ _____ _____

- ☐ A schedule for performance? _____ _____ _____

- ☐ A specific liquidated damages provision? _____ _____ _____

- ☐ A shop drawing schedule? _____ _____ _____

- ☐ Safety (OSHA) responsibilities? _____ _____ _____

- ☐ Shipping or delivery instructions? (FOB location, etc.) _____ _____ _____

- ☐ Method and terms of payments? _____ _____ _____

- ☐ All data required for progress payments? (including lien release, bond, insurance, etc.) _____ _____ _____

- ☐ All data required for final payment? _____ _____ _____

- ☐ Bond requirement? _____ _____ _____

- ☐ Insurance requirements?—Certificates on file? _____ _____ _____

- ☐ Specific or peculiar warranty provision? _____ _____ _____

- ☐ Requirements to provide schedule input? _____ _____ _____

- ☐ Long lead-time items list submitted? _____ _____ _____

- ☐ Oral agreements included in written agreement? (back-up documentation?) _____ _____ _____

- ☐ As-built requirements? _____ _____ _____

- ☐ Taxes (if any) included in price? _____ _____ _____

- ☐ Delivery charges? _____ _____ _____

- ☐ Hoisting requirements? _____ _____ _____

- ☐ Layout requirements? _____ _____ _____

- ☐ Cleanup requirements? _____ _____ _____

- ☐ Protection of work in place and stored materials? _____ _____ _____

- ☐ Union agreements, if required? _____ _____ _____

Figure 7.10

Subcontract Agreement Checklist

	Yes	No	Date
☐ Are subcontract provisions consistent with owner contract (payment, stored materials, schedule, etc.)?	_____	_____	_____
☐ Has the subcontractor been qualified with the owner?	_____	_____	_____
☐ Has a signed copy of the agreement returned to our office without exceptions or conditions?	_____	_____	_____

Figure 7.10 (continued)

Pre-Construction Subcontractor Conference Checklist

The typical pre-construction conference should have the following topics on the agenda:

☐ Safety

☐ The project schedule

☐ Shop drawing schedules and performance standards review

☐ Subcontractors' relationship with the owner and inspectors

☐ The importance of communicating existing and potential problems to the general contractor

☐ The need to maintain proper records, such as daily reports and accident reports

☐ Quality assurance requirements for quality control

☐ Labor concerns, such as jurisdictional disputes

☐ Material handling plan (storage, lifts, cranes, etc.)

☐ Administrative details, including payment procedures, time limits and procedures for requests for estimates and requests for information, and the preparation expected for weekly meetings.

Figure 7.11

Shop Drawings

The general contractor is responsible for compliance with the contract specifications and documents and cannot (as far as the owner is concerned) transfer that responsibility to the subcontractor. In the eyes of the owner, the "buck" stops with the general contractor. It is the general contractor's responsibility to demand that subcontractors submit timely shop drawings that conform to the specifications. Because technological advances or lack of material availability may lead to deviations from the specifications, the general contractor should require the subcontractor to note any necessary deviations (including value engineering changes or methods which may improve the project), and where they occur, in a cover letter attached to the shop drawings. In this way, proposed deviations can be checked for acceptability.

Take the case of the general contractor who rushes to meet deadlines and passes hastily-checked, faulty shop drawings on to the architect/engineer and owner, who also fail to see the error. If it turns out that the error is significant, the general contractor has the expensive responsibility of repairing or replacing the non-spec item. Even though the general contractor may have some consolation in the form of legal recourse against the subcontractor or owner/architect who approved the submittal, the owner may take a negative view toward the general contractor, subjecting him to undue scrutiny for the duration of the job. Furthermore, the subcontractor might lack the financial capability to absorb the error, leaving the general contractor with legal recourse, but an empty wallet.

Records must be kept of all submittals of shop drawings and samples, and should be accompanied by the date of submittal, the requested approval date, and the action taken by the owner and architect/engineer. (See Chapter 5 for Submittal Log.) Records and dates of subsequent actions taken should also be kept if submittals are not approved in a timely manner.

Managing Subcontractors During the Implementation Phase

The general contractor is required to take affirmative, proactive measures in order to get the subcontractors to properly perform their work. The areas that need the most attention are shown in the Subcontractor Management checklist in Figure 7.12; these should be addressed at each weekly subcontractor meeting.

The general contractor has many other responsibilities regarding the subcontractors during the erection of the project. These include: coordination and cooperation, quality assurance, handling backcharges, and managing changes and extras. Throughout the project, the general contractor must be sure to maintain its proper stance between the owner and the subcontractors.

Maintaining the Proper Stance

The contractual relationship shown in Figure 7.1 shows the general contractor in an intermediate position between the owner and the subcontractors. Since this is the usual contractual position of the general contractor, it is clear that he must act as intermediary to maintain control of the project. To stay in the middle, the general contractor must follow the guidelines in the Human Relations Checklist in Figure 7.13.

Subcontractor Management Checklist

The general contractor should ask himself these questions at the start of the job and before the weekly subcontractor meetings.

	Yes	No
1. Has the subcontractor submitted a resource plan indicating work force?	_____	_____
2. Is the subcontractor following the resource plan?	_____	_____
3. Does the subcontractor have a requirement to coordinate its submittals with other equipment suppliers?	_____	_____
a. If so, has the subcontractor properly coordinated?	_____	_____
4. Is the subcontractor responsible for preparing installation drawings or isometrics?	_____	_____
a. If not, should he be asked to do so in order to prevent conflicts?	_____	_____
b. If so, has the subcontractor performed these drawings?	_____	_____
5. Has the subcontractor prepared his plan of work?	_____	_____
a. Is the subcontractor maintaining his sequence in accordance with his plan of work?	_____	_____

   b. If not, why not? _____

_____

   c. If not, what measures should be taken to get the subcontractor to maintain its plan

     of work? _____

_____

d. Is the subcontractor providing input at weekly meetings, thereby enabling the general contractor to adequately coordinate the work with other subcontractors?	_____	_____
6. Does the subcontractor protect its own work?	_____	_____
7. Does the subcontractor adequately protect the work of other craft contractors?	_____	_____
8. Does the subcontractor have a quality control plan?	_____	_____
a. Is the subcontractor checking its work against quality requirements?	_____	_____
b. Are the subcontractor's quality problems holding up other trades?	_____	_____
c. Does the owner's inspector work with the subcontractor through the general contractor?	_____	_____
9. As to Requests for Information:		
a. Does the subcontractor use the architect to do its homework?	_____	_____
b. Are RFI's reasonable and timely?	_____	_____

Figure 7.12

Subcontractor Management Checklist

	Yes	No
10. As to Requests for Estimates by subcontractor:		
a. Does the subcontractor submit proposals in a timely fashion?	_____	_____
b. Are proposals complete?	_____	_____
c. Is impact or schedule delay indicated and justified?	_____	_____
d. Are the subcontractor's prices normally reasonable?	_____	_____
11. As to the subcontractor's work force:		
a. Is it reasonably qualified?	_____	_____
b. Is turnover excessive?	_____	_____
c. Are the subcontractor's supervisory personnel (foremen) appropriately qualified?	_____	_____
12. As to administrative requirements:		
a. Does the subcontractor comply with written notice requirements?	_____	_____
b. Are payment requests accurate and timely?	_____	_____
c. Is the subcontractor submitting required data (as-builts, etc.) for contract close-out?	_____	_____
13. As to acceptance of work:		
a. Is the subcontractor participating in punch list development?	_____	_____
b. Is the subcontractor working off punch lists on a timely basis?	_____	_____

Figure 7.12 (continued)

General Contractor Human Relations Checklist

To properly maintain his contractual role, the general contractor must follow these basic rules:

☐ Do not sign off completely on change orders unless affected subcontractors have also signed and all are satisfied with the work involved.

☐ Get owner's interpretation and directive on scope of work disputes and then pass them on to subcontractors.

☐ Always pass owner's directives (regarding acceleration, manpower, or overtime) on to subcontractors if the source is an act or omission of the owner.

☐ Give subcontractors a chance to present their views on inspection disputes. Discuss the situation with the owner and get direction for the corrective work. Then, direct the subcontractor per the owner's directive.

☐ Do not allow the owner to deal directly with the subcontractors.

Figure 7.13

The Duty to Coordinate and Cooperate

Subcontractors are typically hired by the general contractor to perform various craft portions of a construction project. The general contractor must sequence the schedule and coordinate progress in order to enable each subcontractor to work efficiently and without unreasonable interference from either the general or other subcontractors. Therefore, it is a vital duty of the general contractor to direct the work, since the general contractor is normally the only entity that has a direct contractual relationship with all of the subcontractors.

Just as the general contractor should expect to do its job without undue interference from the owner, so the subcontractor is justified in expecting no unreasonable interference with its work. The general contractor who fails to take on the responsibility of coordinating and ensuring cooperation opens the door to claims for a subcontractor's increased expense resulting from that failure.

To ensure cooperation, the general contractor must exert reasonable efforts to cause every preceding activity to be properly performed and all appropriate preparations to be made when each subcontractor moves into an area. This is what we call a "state of readiness," which can be achieved by following the Site Preparation Checklist in Figure 7.14. Figure 7.15 is an example of a simple resourced network that can be used for each work area.

As part of the responsibility to coordinate and cooperate, the general contractor may sometimes have to pressure a subcontractor who is behind schedule to catch up. While it may be clear that a particular subcontractor is behind, the fault may lie elsewhere. So, the question arises: "What *does* the general contractor do when a subcontractor is behind schedule?" Figure 7.16 is a flow chart of actions that should be taken. A sample letter that can be used by the owner or the subcontractor is shown in Figure 7.17.

Quality Assurance

The need to check subcontractors' work cannot be emphasized enough. If the subcontractor erects a portion of the job which later fails and causes injury or other damage, the general contractor may be held liable. In cases where the general contractor was unable to establish proof that the subcontractor's work complied with contract requirements, several lawsuits have been decided against the general contractor. The financial exposure resulting from a lack of quality assurance could outstrip the limits of the general contractor's liability insurance. Furthermore, it could damage the general contractor's reputation among owners.

The general contractor who escapes such dire consequences may still wind up footing the bill for rework or paying for an extension of the job.

Clearly, it is up to the general contractor to set up its own quality assurance plan and to implement the proper quality control procedures to ensure that the subcontractor's work definitely meets the contract requirements. The subcontract agreement may allow the general contractor to take over work by any subcontractor who is making unsatisfactory progress after three days written notice to that effect. When the general contractor

Site Preparation Checklist

Part of managing the coordination of craft contractors is getting a work area ready. This means that part of the general contractor's job is to exert a reasonable effort on behalf of the subcontractor to:

☐ Provide access to the workplace.

☐ Provide the actual work space.

☐ Provide work space that is ready for the work to begin. In other words:

    _____ that the work space (such as a concrete slab for a drywall contractor) is complete and meets the correct tolerances to accept the work.

    _____ that any equipment or materials furnished by others (such as fan coil units in the case of the drywall contractor) are delivered and installed on a timely basis.

    _____ that environmental conditions (such as temperature, sealing in building, etc.) are in accordance with accepted standards.

    _____ that other trades are not working in conflict with that subcontractor.

Using a simple resourced network for each work area (such as shown in the example in Figure 7.15), the general contractor has an effective tool for coordinating the various craft contractors.

Figure 7.14

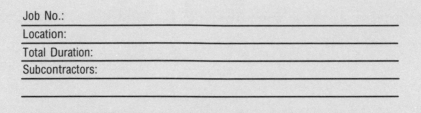

Description	Man-hours	Crew Days
Layout		
Top Track		
Frame Walls		
Set Door Frames		
Board 1 Side		
Elec. in Walls		
Mech. in Walls		
Insulate		
Backing		
Board Side 2		
Fire Tape		
Frame Clg.		
Elec. in Clg.		
Mech. in Clg.		
Hang Clg.		
Tape Clg.		
Acoustical Spray		
Clean-up		

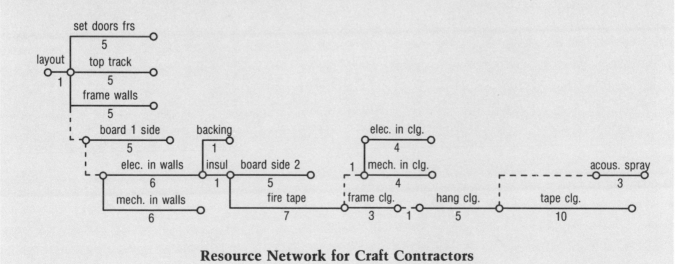

Resource Network for Craft Contractors

Figure 7.15

Schedule Review and Corrective Measures Checklist

1. Cause for review: _____
   _____
   _____
   _____
   _____

2. Subcontractor appears to be behind schedule. When compared to schedule,
   subcontractor performance is: _____
   _____
   _____

3. Cause of the delay identified as: _____
   _____
   _____

4. Responsible party identified as: _____
   _____
   _____

5. Take action as follows:

   ☐ If owner is responsible, forward owner's directive to subcontractor. (See sample letter in Figure 7.16.)

   ☐ If subcontractor is responsible, issue a directive to the subcontractor to meet the schedule. (See Figure 7.15)

   ☐ If contractor is responsible, negotiate with subcontractor, but expect to pay costs of overtime, etc.

Figure 7.16

Dear Mr. Subcontractor:

According to the latest construction schedule to which you committed at our weekly subcontractors' meeting on (Date), the following is a list of activities that you have either not started or have not completed within the construction schedule.

<u>Activity</u>   <u>Projected Slippage to Date</u>

(This applies where there is an Owner directive or constructive acceleration.)

The Owner (has directed us to complete the project within the contract schedule), (has threatened liquidated damages if we do not meet the schedule). Therefore, in accordance with the Owners' directive, it is requested that you respond in writing by (Date) with the affirmative steps that you intend to take to make up the lost time and to comply with the Owner's demand to meet the schedule. In the meantime, you should begin implementing the steps necessary to meet the Owner's schedule.

(Use this paragraph if you believe the subcontractor is at fault.)

On (Date), you provided us with a projected manpower loading for this project. Our daily records indicate that you have consistently undermanned the project (see attached manpower summary). You are directed to provide a written statement of your recovery plan, including all steps you will take to catch up with the schedule. You are expected to commit to this plan and to immediately begin its implementation. This notification is given pursuant to Clause _____ of the subcontract agreement.

Figure 7.17

assumes such work, extra costs must be documented and should be recorded using a separate cost code. The daily construction report to the home office should include statements of no work or unsatisfactory work by subcontractors.

Backcharges

The general contractor should, as a matter of policy, strive for a firm, but cooperative relationship with subcontractors. Those general contractors who are unreasonably demanding of subcontractors will eventually feel the adverse effects of these actions. Part of a reasonable, cooperative approach is to refrain from making a practice of backcharges for the sake of backcharges. Where a real backcharge item does exist, the cost should be properly documented, and the necessity for the backcharge communicated to the subcontractor. It is also important that the general contractor give the subcontractor prior written notice of the backcharge. When issuing a backcharge, the general contractor should assess the impact that the subcontractor's action (or lack of action) had on the overall schedule.

Changes and Extras

All changes and extra work that the general contractor directs the subcontractor to execute should be in the form of a written Work Authorization. When verbal instructions are used in lieu of written directives, the general contractor quickly loses control of a project. Parties may forget the sources and/or reasons for the changes if these are not documented properly. The scope of the work should be clearly stated in the work authorization, including man-hours, materials, and equipment. When time allows, the general contractor should negotiate a lump sum price for the extra work before it is performed.

Work authorization directives are issued to subcontractors for three reasons:

- oversights in the initial subcontract agreement
- change orders issued by the owner
- on-site incidents, such as storms, accidents, etc.

The general contractor must prompt the subcontractor to file claims for extras resulting from the owner's change orders within the time limit specified in the contract documents. The general contractor may even have to assist the subcontractor in preparing the claim so that the general can submit it to the owner in a proper and timely manner. A short training session with the subcontractors regarding administrative procedures and requirements can go a long way to alleviate or prevent many disputes.

The general contractor should either include the subcontractor in change order negotiations with the owner, or should procure both the supporting data and the authorization to negotiate from the subcontractor. The limits of this authority to negotiate should be well established at the start. The general contractor who settles for less than the amount that the subcontractor had requested, without the subcontractor's consent, may be liable for the difference.

When the work authorization reflects necessary work that was left out of the subcontract agreement, the subcontractor enjoys a strong bargaining position. Nevertheless, the general contractor should always negotiate for a reasonable price before allowing the subcontractor to proceed. If the subcontractor will not agree to a reasonable price, the general contractor should have an alternate source available.

Project Close-out

The final months of the project are often the most difficult to manage. By this point, original plans have been largely executed; crafts are usually crowded into the last areas of work; materials may be in short supply, and tools and equipment are in need of repair or replacement or are in demand elsewhere. In a way, the close-out phase of a project must be managed as if it were a new job just beginning. To properly manage subcontractors during close-out, the general contractor should follow the guidelines in the Project Close-out Checklist in Figure 7.18.

Subcontractor Evaluation

The performance of the subcontractors should be reviewed periodically during the project. The general contractor must have an objective assessment of the subcontractor's work to use as a negotiating tool for upgrading a poor subcontractor effort. A record of such evaluations can also be valuable in the selection of subcontractors for future projects. Relying on the "gut feelings" of the superintendent or project manager may place too great of an emphasis on personal observations or increasingly distant or selective memories. Answering the questions and including comments in the Subcontractor Evaluation checklist (Figure 7.19) will give the general contractor the information needed to objectively assess the subcontractor's work.

Summary and Conclusions

The operative word in subcontractor management is *management*, which includes:
- selecting qualified craft contractors
- establishing a clear scope of work
- establishing a schedule for administrative duties, such as submittals
- creating a team concept
- establishing and updating a construction schedule
- knowing and doing something about the risks associated with a particular subcontractor
- monitoring the work
- coordinating the work of all the trade contractors
- demanding routine contract compliance
- treating subcontractors fairly

Remember, most legal problems on construction projects follow a management failure by someone—don't let it be you!

Project Close-out Checklist

To properly manage subcontractors during project close-out, the general contractor should:

☐ Update the schedule often and re-plan the final months, bringing the subcontractors into the scheduling process.

☐ Prepare and oversee completion of a preliminary punch list before each subcontractor leaves the project.

☐ Send notification to subcontractors informing them of data requirements (operating or maintenance manuals, brochures, as-built drawings, warranties, waivers of lien, etc.), and that final payment will be withheld pending submittal of that data.

☐ Tie up all administrative "loose ends." Backcharges and change orders should be negotiated while key people are still available in body and spirit, in other words, while the details are fresh in their minds.

Figure 7.18

Subcontractor Evaluation Checklist

	Yes	No
1. Are excessive backcharges being assessed?	_____	_____
2. Did subcontractor monitor quality control?	_____	_____
3. Did subcontractor provide input to schedule?	_____	_____
4. Were shop drawings submitted per schedule?	_____	_____
5. Were materials being delivered per schedule?	_____	_____
6. Are materials checked (quantity/quality) on arrival?	_____	_____
7. Did work progress per schedule?	_____	_____
8. Were punch list items completed in a timely manner?	_____	_____
9. Did subcontractor support schedule updates?	_____	_____
10. Are there unresolved claims?	_____	_____

Figure 7.19

Chapter Eight

Changes Delays, and Acceleration

Chapter Eight

Changes, Delays, and Acceleration

It is rare indeed that a construction contract does not undergo some change due to unanticipated conditions or a change in plans. The contractor is exposed to three major risks associated with change:

- *Scope of Work* increases or decreases
- *Differing Site Conditions*
- *Time Changes* in the form of:
  Delays (extensions)
  Suspensions (shop work)
  Acceleration
  Sequence or production flow changes

No matter how well the contractor priced the job or how efficiently the work is produced, his profit can be eroded if he either fails to protect his contractual rights, or is unable to effectively price and negotiate changes or delays. The contractor must be able to:

- *identify* a priceable change or delay
- provide timely written *notice*
- document its *cost* effect
- document its *schedule effect*
- *prove it* to the owner

He must be able to carry out these tasks with the same proficiency as placing concrete or pulling wire. Today's contractor must be a contract manager as well as a builder.

As emphasized in Chapter 4, being a contract manager begins with reading the contract (RTC). The changes, Suspension of Work, Changed Conditions and Delay clauses (or the Default clause in federal contracts) are the key contractual mechanisms for determining whether the contractor is entitled to additional compensation for changes or delays. There are other clauses that may pertain as well (e.g., variation in quantities, etc.). However, the main point is this: Read *your* contract and become familiar with its particular requirements. Become familiar with the principles of contract interpretation presented in Chapter 4. Two other important points should be made at the outset of this chapter:

1. **Written notices** are always required by clauses that may result in the owner having to pay the contractor additional money or grant the contractor additional time. Abide by these documentation requirements; do not count on verbal discussions to protect your rights. This does not mean getting involved in a letter-writing "war." It does mean that if your contract requires written notice of a change, comply with the contract or be prepared to lose what could otherwise be a valid claim.

2. **Document** as you go along. "Real time" documentation by field personnel is better than any expert witness. Documentation will be discussed further in this chapter, but for now, remember that an extra cost incurred but not properly documented may be your loss, even if you may have an otherwise valid claim. Remember Rule #10 in Chapter 4 on Contract Interpretation; the necessity to document has been elevated to a *rule* of contract interpretation.

The purpose of this chapter and the checklists set forth in it is to assist the contractor in protecting its profit in the face of changes, differing site conditions, or time impact. Again, please read and use this chapter within the framework of the rules of contract interpretation set forth in Chapter 4, and study the unique features of your specific contract.

Scope of Work Changes

The most frequently used clause in a construction contract is the *Changes clause*. The Changes clause was developed to enable an owner to change his/her mind, add quantity or quality, or correct a mistake. Generally, the owner can unilaterally direct a change and the contractor must perform the work, even if price has not yet been agreed upon. In this way, a change will not stop the work. (There are a few changes clauses which require bilateral agreement; these changes clauses are often found in municipal works contracts, so be on the lookout for them.)

There are two types of changes. One is a **formal change**; the other is a **constructive change**. A formal change is issued by an authorized party. That party indicates a scope of work change, directs the contractor to perform the extra or changed work, and negotiates an equitable adjustment (price). In other words, everything is settled except *how much*. There is no question that an equitable adjustment is due. The formal changes are the easy ones, since the owner is admitting financial responsibility for the extra or changed work.

Constructive Changes

A constructive change is the result of a dispute in *interpretation*. Generally, the owner initially does not admit that the extra or changed work is his financial responsibility. Examples of a *Constructive Change* are:

A. Holding a contractor to a higher standard of performance than the contract calls for. For example:

The contract specification describes a concrete 1/4" in 10' for slab on grade. The contractor places the slab on grade at a tolerance of 1/2" in 10'. The inspector rejects the work and requires corrective action. If the contractor can demonstrate that by industry standard (referring to referenced ASTM'S and ACI'S), a 1/4" deviation to a 1/4" tolerance (i.e., 1/2" in 10') is permitted, then the contractor's interpretation will prevail and he will be entitled to an equitable adjustment for grinding the SOG (slab on grade) down to 1/4" in 10'; (assuming the contractor gives timely written notice that he considers the inspector's directive to be a change to the contract).

B. Causing work to be inspected to higher standards than specified, as in the following example.

Assume the technical specifications require visual inspection of welds, but the resident engineer directs the contractor to use dye-penetrant, radiographic inspection. Again, assuming timely written notice, the contractor may be entitled to additional cost for these more involved inspection methods.

C. Interpretation disputes over ambiguous specification language. To illustrate:

Assume the technical specifications provide for "a perfect weld," or "non-porous concrete," or "completely even." What do these terms mean? If the owner directs the contractor to perform in accordance with a standard that exceeds a level consistent with trade practice (and again assuming timely written notice), the contractor may have a claim for extra costs as a result of compliance with this directive. We would term this directive, then, a "constructive change," or a change in the interpretation of the documents or acts of the parties.

D. Compliance (or attempted compliance) with a defective specification. For example:

The contractor attempts to comply with a specification that turns out to be defective. For example, during pile driving, the contractor experiences a high breakage rate of piles because the dowel overlap connecting the mechanical splice was designed too short. The contractor's rate of production suffers substantially until the deficiency is corrected. With timely written notice, the contractor's efforts to comply with the defective specification would be compensable.

A contractor must know the rules of contract interpretation discussed in Chapter 4, and carry out all documentation requirements. The contractor must also know how to manage the changes clause. The checklist in Figure 8.1 contains key words from a standard changes clause and provides the meaning and the appropriate contractor response to those words.

The format of the letter of notification for constructive changes is shown in Figure 8.2. Explanations of each step are provided in the left margin, in parentheses.

Changed Conditions

There are two types of changed conditions which are described below as Clause I and Clause II.

Clause I

The first type of frequently occurring changed condition involves subsurface or other conditions at the site that *differ materially from those indicated in the contract.* Compensation therefore depends upon encountering a condition that differs materially from that which was indicated by the contract documents. The actual conditions discovered need not be *contrary* to those specifically stated in the contract documents; it is enough that the documents contain "indications" which would lead a contractor to reasonably infer that he would not encounter the conditions that were, in fact, encountered.

For example, if the log borings indicate clay soil and the contractor discovers ledges of rock not shown in the borings or the soils report, the contractor is entitled to a "Clause I" differing site condition. Some common types of changed conditions are listed in Figure 8.3.

To recover for this type of changed condition, i.e., one which is, under AIA language, "at variance with conditions indicated by the Contract Documents," a contractor must generally show four things:

1. the actual condition
2. the conditions indicated by the plans, specs, and other contract documents

Standard Changes Clause: Terminology and Contractor Responses		
The Term or Phrase	What It Means	What Contractor Should Do
The Owner	Only the owner or his authorized representative may order changes.	
Authorized Representative	The party (or parties) designated by the owner to make project decisions and to authorize changes on behalf of the owner. Even the architect, unless so designated by the owner, cannot authorize changes.	Find out who is authorized. This is especially important on any government (federal, state, or municipal) project.
Inspector	The inspector usually cannot order extra work. If he tries, it should be confirmed in writing with the owner.	Be aware of inspectors and careful of the extent of their involvement.
Lending Agency	On some projects, the lending agency must be notified of changes and must concur.	Be sure to comply with the provisions stated on the financing agreement if it is made part of the contract. Keep the lending agency informed, and document all exchanges of information.
Written notice	Notice must be issued in *written form*.	This is essential. If the owner orally directs work, immediatley confirm it in writing.
WRITTEN ORDER: Formal Changes	There are *formal* changes and constructive changes. A formal change is one that complies with the formalities of the Changes Clause (i.e., the owner requests that the contractor submit a proposal for a change to the contract).	The contractor prepares a cost proposal based on the owner's changes, submits it to the owner, and negotiates it. The owner then directs that the work be performed or not performed.

Figure 8.1

Standard Changes Clause: Terminology and Contractor Responses

The Term or Phrase	What It Means	What Contractor Should Do
Unilateral Right to Order	The owner may alternately unilaterally direct the contractor to perform the changed work, even before it has been priced. The contractor *must* comply with the order. Note: Some contracts provide for a bilateral agreement before a change can be authorized. (This is especially true of Municipal contracts.)	The contractor must comply with a unilateral change order. The only exception is if the change is a *cardinal* change, that is, not in the scope of the contract. The contractor should immediately confirm the directed change of the scope of work in writing.
CONSTRUCTIVE CHANGES Interpretation (see Chapter 4 for more on interpretation)	Even though the owner and the contractor disagree on the interpretation of the contract, the owner may direct the contractor to abide by the owner's interpretation.	Immediately confirm all directives in writing.
Over-inspection	An inspector may hold the contractor to a higher standard than the contract specifies. Or, he may unreasonably reject the contractor's work as nonconforming.	Document all decisions. Direct all concerns to the owner and/or architect immediately.
Congestion	The design is simply not constructible because of limitations of space. For instance, the piping will not fit into the chase, a fact that could not have been determined from single or double line drawings on the prints.	*If* the contractor gives written notice on a timely basis, and if the contractor's interpretation is correct, these circumstances give rise to what is called a *constructive change.* Ultimately, the contractor (if his position is correct) will be paid just as if this were a formal change.
General Scope Furtherance of Design	General scope means "in furtherance of the design." The owner has the right to order changes in the general scope of the contract, whether or not the contractor agrees, or whether the price has been agreed upon in advance.	Document all changes immediately.
Cardinal Changes	A cardinal change is outside the scope of the contract, like adding another wing to the project. The owner does not have the right to order—unilaterally—a cardinal change.	On such major changes, the contactor is protected from being unilaterally forced into doing the work before a price is agreed upon.
CLAIM FOR ADDITIONAL WORK Time of Written Notice	The contractor must provide written notice within the time stated in the contract.	Document all costs associated with the additional work that constitute a claim.
Prior to Working or Incurring Cost	Before proceeding with the work. The reason for this provision is to give the owner the opportunity to mitigate his damages.	Notice requirements are being more stringently enforced. Professional administration of the contract demands that timely, written notifications be made; otherwise, the contractor will be performing at his own risk.

Figure 8.1 (continued)

_____ Date

RE:

Dear Mr. Contract Officer,

(Establish basis of bid)	1.	Our company bid on the following contract requirements:
(Contract reference)		Specify applicable specs, drawings, general conditions.
(Contractor's interpretation)	2.	What this meant was: Describe quantity, quality, method of or space for installation. In other words, how this section of the contract was priced.
(How this is being changed)	3.	You (name who, refer to directive or other source) have requested (or directed) that the following changes (additions) be made:
		Describe the nature of the change and how it is different from your interpretation of the contract.
	4.	Thus, the difference between the original contract and this change (directive) is:
(Describe actual, quantifiable, difference)		Quantity: Quality: Methodology: Time zone for performing:
(Substantiate your basis of interpretation)	5.	The basis for your interpretation of the contract is: a. Basis of estimate. (You aren't paying twice for the same thing.) b. Rules of interpretation applied to this issue. c. Industry standard or practice.
(When authorization is needed)	6.	We would appreciate your immediate authorization by (specify date) to proceed with this work as an extra to the contract.
(Reserve rights if necessary)	7.	We reserve the right to all costs associated with the change (including the impact on other work) when those costs can be determined.

Figure 8.2

3. the variance
4. that proper notice of that variance has been given

The most critical element is usually the "conditions indicated" in the contract. If the contract does not address these conditions, there can be no recovery, because there was no representation or "indication" made by the contract documents. It is not necessary that the contract indications be explicit or specific, as long as they are adequate to inform a reasonable bidder not to expect the adverse conditions.

"Indications" may also be inferred from a reading of the contract documents as a whole. This means that the design or construction procedures specified in the contract reasonably imply that a certain type of condition is in effect "indicated," but they do not expressly state that condition "X" will or will not be encountered. (For example, a changed condition was found in one case where soils were physically incapable of being compacted in accordance with a specified compaction test.) Thus, the performance standards, the design, or the materials specified in a contract may provide a sufficient "indication" of the site conditions to warrant recovery when *actual* conditions prove to be different.

Clause II

A second type of changed condition is an unknown physical condition of an unusual nature, differing materially from those *ordinarily encountered and generally recognized as inherent in work of the character called for in the contract*. This second type of changed condition is less frequently alleged and is more difficult to prove.

In Clause II type changed conditions, recovery does not depend on comparing contract documents with conditions actually encountered. However, if the contractor should have reasonably anticipated the condition (notwithstanding the fact that it is not mentioned in the contract documents), then it will be regarded as an *unknown* or *unusual condition*. The question is always

---

Common Types of Clause I Change Conditions

Conditions Different from Those Indicated in the Contract

- Encountering rock or other material in an excavation area when such material was not indicated based on the logs of the exploratory borings which had been made available to bidders and were a part of the contract.
- The discovery of boulders larger than indicated by the soil test borings which were incorporated into the contract.
- Encountering substantially more rock in the excavation area than that indicated by the logs or borings.
- The discovery of ground water at an unforeseeably higher level.

Figure 8.3

whether the bidder's judgment and work interpretation was *reasonable* at the time of bidding. In making its decision on the contractor's right to recover, a court will consider such factors as common knowledge and custom in the industry, manufacturer's instructions and recommendations, and traditional assumptions ordinarily involved in bidding a particular type of work.

For example:

> *The log borings for a project might indicate the extensive presence of "metagraywacke," a type of rock which the contractor knows requires blasting. The log borings also show the presence of some water in the area. The contractor anticipates drilling and shooting, and machine excavating the rubble in the trench. When the metagraywacke is blasted and exposed to the water, it loses its rock characteristics, and becomes a dense clay slurry which cannot be pumped out of the trench. Such an unexpected and unforeseeable condition would be a Clause II changed condition.*

Figure 8.4 is a flow chart of events that may lead to a contractor's successful claim for damages under either type of Differing Site Condition Clause.

Excluded Risks

Finally, certain types of conditions occur frequently in construction contracts, but are not considered to fall within the category addressed by the Changed Condition clause. Some examples of these excluded risks are shown in Figure 8.5.

Site Investigation Clause— Disclaimer Clauses

Frequently, construction contracts contain site investigation clauses. These can make recovery under a changed condition clause more difficult by preventing a contractor from successfully arguing that actual conditions were different from those that were or should have been anticipated. However, the contractual requirement that the contractor investigate the site does not void *per se* the changed conditions clause of the contract.

AIA document A201 provides as follows regarding site investigation:

2.2: "Each bidder, by making his bid, represents that he has visited the site and familiarized himself with the local conditions under which the work is to be performed."

The AIA General Conditions also provide:

1.1.2: "By executing this Contract, the Contractor represents that it has visited the site, familiarized itself with the local conditions under which the work is to be performed, and correlated its observations with the requirements of the Contract documents."

In the case of federal contracts, a typical site investigation clause generally requires only a reasonable investigation—information that a reasonable, experienced, and observant contractor could discover, rather that what a trained geologist or other specialized expert might be able to discover. Nor does the clause create a requirement that the contractor conduct independent technical investigations, such as obtaining subsurface boring and core samples when the bid documents already provide such information.

In both state and federal contracts, when the contract documents specifically represent that a certain condition exists, one can rely on that representation. However, if the contract documents do not specify the conditions that may be encountered, the contractor may be obligated to inquire whether the owner or architect has relevant information in its possession.

Related Checklists and Documentation

The Pre-bid Site Investigations Checklist (introduced in Chapter 2) and the Differing Site Conditions Checklist (shown in Chapter 4) will guide the contractor through the important details and provide a documented trail to aid in managing changed conditions. In addition, photographs, memoranda, and minutes of

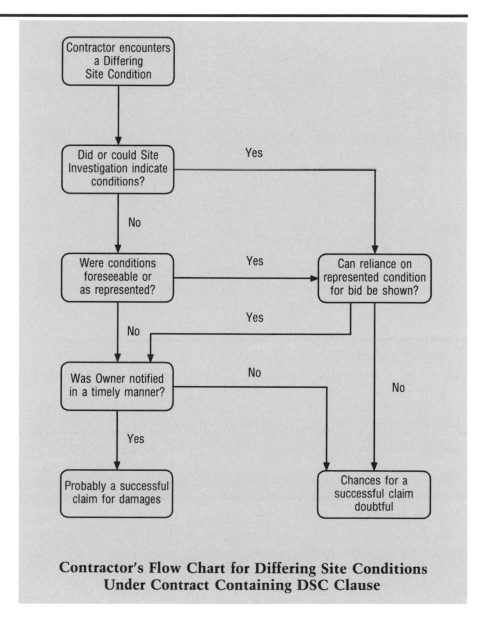

Contractor's Flow Chart for Differing Site Conditions Under Contract Containing DSC Clause

Figure 8.4

meetings with owners should be maintained in the bid file. Any information pertaining to physical data that was collected for the bid file should be retained and reviewed.

Figure 8.6 is a summary of the risks assumed by the owner and contractor under the Site Investigation Clause and Differing Site Conditions.

The Differing Site Conditions Checklist in Figure 8.7 provides a blueprint for gathering the information necessary to establish a differing site condition.

Time: Delays, Suspensions, and Accelerations

The issuance of a Notice to Proceed starts the clock—in what can be a race against time. The contractor's profit is tied up in how well time is managed (including managing the interferences with the contractor's use of time). No matter how well planned a job is, there are a host of daily factors which challenge the contractor's management of time. Some of those factors are the contractor's own responsibility (or his subcontractors'), some are acts of God (such as weather), and some are interferences caused by the owner and/or his/her representative. Some of these delays or suspensions are compensable; some are not. Figure 8.8 lists the various causes of delays and classifies them by compensability to the contractor. Remember—examine your own contract for its unique features.

Suspension of Work

Not all contracts have Suspension of Work clauses, but most do. The Suspension of Work clause, like the Changes clause, contemplates both Formal and Constructive Suspensions of Work.

A formal suspension is a written order from the owner for the suspension or interruption of the work. The contractor is entitled to *all* costs associated with any delay that such an order may *cause*, if *timely written notice* is given to the owner. Under most Suspensions of Work clauses, the contractor is not, however, entitled to profit from the suspension, but should recover only his losses.

Risks Excluded from Changed Conditions Clauses

- Weather conditions will not ordinarily constitute a compensable changed condition unless they combine with another condition which meets the requirements of the clause to result in a loss.
- Changes in wage rates during contract performance or increases in the costs of materials.
- Inability to obtain materials.
- Foreseeable overruns in contract quantities.

Figure 8.5

In addition to a formal suspension of work, the owner or his consultant may also stop or suspend a contractor's work by their *actions*, resulting in what is termed a *Constructive Suspension of Work*.

The following are examples of Constructive Suspensions:

    a. An unreasonable period of time in approving shop drawings, models, samples, etc.

    b. An unreasonable period of time in investigating a changed condition and issuing direction or a contract modification.

    c. Unreasonable denial of access to job site.

    d. Unreasonable inspection, delay in observing tests, etc.

    e. Any other unreasonable period of time in issuing direction, decisions, or other responses by the owner.

Of course, the contractor must *preserve its rights* through *timely written notification*. Also note that many Suspension of Work clauses do not permit the contractor to receive profit on the cost incurred during the suspension.

Acceleration

Formal Acceleration

Under most contracts, an owner may formally direct an acceleration (e.g., "The contractor is directed to complete Phase I

Owner's and Contractor's Risks Under the Site Investigation and Differing Site Conditions Clauses	
Pre-Bid Site Investigation Clause	
Responsibilities imposed on the contractor <ul><li>To visit the site</li><li>To observe and record essential information</li><li>To review specs/drawings</li><li>To know the labor market, non-physical conditions</li><li>To know the weather conditions</li></ul>	**Obligations not imposed on the contractor** <ul><li>To assume misrepresentation</li><li>To have expert knowledge</li><li>To perform a full soils investigation</li></ul>
Differing Site Condition Clause	
Clause I **Owner Risks** <ul><li>Misrepresentation</li><li>Spearin application</li><li>Concealment</li><li>Failure to issue direction</li></ul> **Contractor Risks** <ul><li>What he knew or should have known</li><li>Illogical or incomplete inferences</li><li>Written notice on a timely basis</li><li>Proof of changed condition</li><li>Duty to mitigate damages</li></ul>	**Clause II** **Owner Risks** <ul><li>Unusual</li><li>Unforeseeable</li></ul> **Contractor Risks** <ul><li>Bound by knowledge of previous experience in area</li><li>Written notice on a timely basis</li><li>Proof of unusual or unforeseen condition</li></ul>

Figure 8.6

Differing Site Conditions Checklist

☐ Read the differing site condition clause of the contract and follow its procedural requirements.

☐ Give IMMEDIATE and WRITTEN notification to the owner and ask for direction!

☐ Take PICTURES.

☐ Get the owner's representative to the site immediately!

☐ If in doubt, get a soils engineer or other consultant to the site immediately.

☐ Set up SEPARATE COST CODE and track direct costs associated with the extra work. Daily diaries/logs show the extra work being performed each day, the crew, equipment, overtime, etc.

☐ Review the IMPACT on the SCHEDULE to determine what additional activities are being affected.

☐ Attempt to develop an analysis of the additional time required and schedule on an optimum rather than a crash basis.

☐ If no or inadequate time is permitted by owner, DOCUMENT the EXTRA COST according to procedure on acceleration.

☐ Remember, most differing site condition claims arise during the crucial first 25% of the job; this is where a change in sequence or momentum can have a devastating effect on the job, so the changed condition must be really well managed!!

☐ For conditions which differ materially from those indicated in the contract:

    ____ Show what the actual conditions are.

    ____ Show what conditions were indicated by the plans, specs, and other contract documents.

    ____ Show the difference between indicated and actual conditions.

    ____ Put owner on notice of changed conditions.

☐ For unknown physical conditions of an unusual nature differing materially from those *ordinarily encountered*:

    ____ Show that the contractor should not have reasonably anticipated the condition.

    ____ Show that the pre-bid site investigation was adequate and reasonable.

    ____ Show what the bid anticipated with regard to this condition and how it was "covered."

☐ Bring in an expert early to identify differing conditions and propose possible alternatives.

☐ Prepare list of options for owner. Get owner's direction.

Figure 8.7

of the project thirty days earlier than the contract schedule.") In such a case, the contractor provides immediate written notice of the claim and will generally be entitled to collect the additional costs associated with meeting the accelerated performance date.

Constructive Acceleration

Constructive Acceleration is the most common type of acceleration, and yet it presents a high risk to both contractor and owner. The elements of a Constructive Acceleration are listed below. The Constructive Acceleration process is shown graphically in Figure 8.9.

- Contractor is entitled to an **excusable delay**.
- Contractor notifies owner and asks for additional time.
- Contractor demonstrates that additional time is justified.
- Owner **denies** request for additional time; holds contractor to original schedule.
- Contractor notifies owner that he considers this an acceleration.

Criteria for Three Types of Delays and Compensability to the Contractor

Time Extension only in the case of . . .

• Acts of God, e.g., natural disasters such as tornadoes, floods, or earthquakes. • Public enemy, e.g., terrorist destruction of work. • Sovereign acts of the government, such as embargoes, price controls, military operations, etc.	• Unusually severe weather (must prove with official records) • Labor strikes (must be unforeseeable)

Time and Money in the event of . . .

• Change Orders • Differing site conditions • Late or deficient owner-furnished property • Owner's failure to make timely payment to contractor (damages limited to time plus interest) • Owner's delay in approving shop drawings, submittals, samples, or schedules	• Improper inspection or failure to make inspection by owner • Defective specifications • Unusually severe weather compounded by act of owner, such as delay • Any other failure by owner to perform contractual duties

No Relief in case of . . .

• Labor shortages	• Financial difficulties by prime or subcontractors (entitled to time extension if owner has failed to make timely payment)

Note: Regarding No Damages for Delay Clause:
Many contracts contain a No Damages for Delay clause which provides that the contractor is entitled to time relief only and no compensation for these delays. Although it may appear that these clauses are inherently unfair, courts hold them to be enforceable, but with many exceptions, including:

- accelerations
- unreasonable delays
- breach of contract

Please review the checklists in Chapter 4 and seek counsel on individual problems concerning No Damages for Delay.

Figure 8.8

- Contractor responds to acceleration by working overtime, adding personnel, etc.
- Remember, if the contractor has **also delayed the project,** it may have concurrent causes and **no** grounds for acceleration.

Accelerations can cause a "crash configuration," such as shown in Figure 8.10. When a contractor's labor forces are in a crash mode, its costs can go up on an exponential scale, due to the effects of oversized crews, crowding, and overtime on labor productivity.

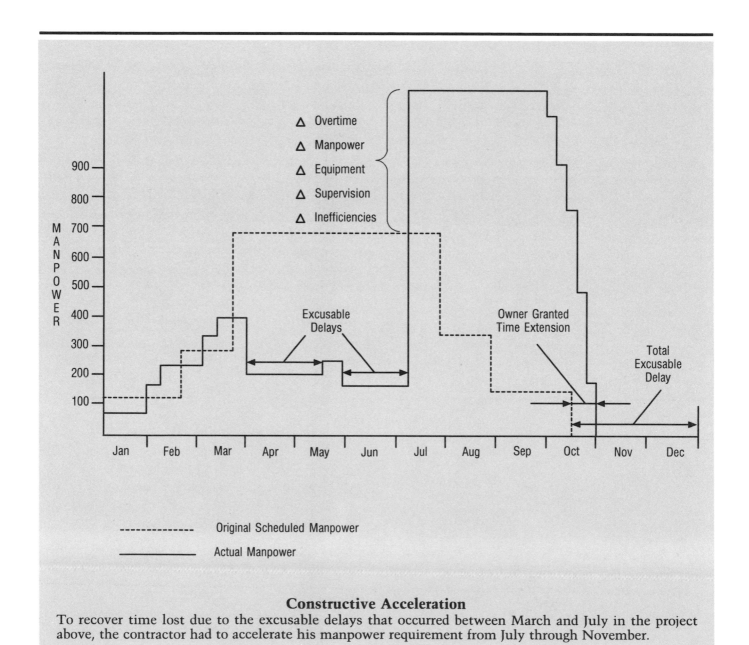

Constructive Acceleration

To recover time lost due to the excusable delays that occurred between March and July in the project above, the contractor had to accelerate his manpower requirement from July through November.

Figure 8.9

236

"Clean Hands" and Causal Relationship

Several words of caution are appropriate at this point. First, in order to obtain an extension of time or to have a compensable Constructive Acceleration, the contractor must demonstrate *"clean hands,"* i.e., that this problem occurred without the contractor's fault or negligence. Second, in order to obtain relief or be entitled to a remedy, the contractor must demonstrate a causal relationship, i.e., that he was, in fact, damaged in some way by the owner's action.

Pricing

The concept of pricing is generally expressed in terms of "equitable adjustment," or placing a contractor in a "whole" position so that a change or owner-caused delay will not degrade the contractor's financial position. The following procedure sets forth the steps in pricing a change to the contract.

Changed and Unchanged Work

It is vital to realize at the outset that a change can affect not only the changed work itself, but also the technically "unchanged" work. The *changed work* is that work necessary to carry out the directions shown on the revised drawings and/or specification. For example, if a drawing adds a light fixture, the changed work is that light fixture, including the conduit, wiring and other associated materials and labor necessary to install one light fixture.

The unchanged work and the crews performing it may also be affected by the changed work. For example, if a revised drawing adds a light fixture, and the affected electrical crew works on the

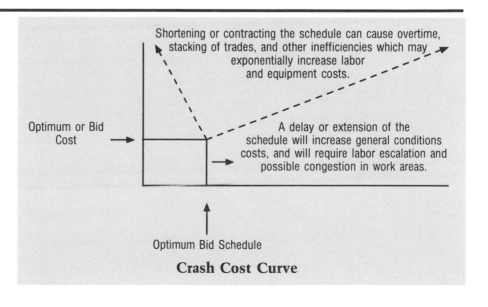

Figure 8.10

Crash Cost Curve

ceiling at the same time as the drywall is being installed, the drywallers' work may suffer from interference. The ceiling grid may also need reworking, and so on.

The Impact Identification Checklist shown at the end of this chapter provides an organized means of identifying the cost item in the changed and unchanged work.

Pricing the Changed Work

In pricing the changed work, the first step is to "scope out the work" itself. This means that the contractor must:

- Review the request for proposal or change itself.
- List the drawings affected and red-line the changes to the drawings.
- Review the specifications for changes.
- Send copies of the RFP to all subcontractors and affected suppliers, with a requirement to submit a price by a given date.
- Now, review the extra work which must be done.
  Will rework be necessary?
  Will tear-out or finishing work be required?
  Will additional *manpower* be required?
  Will additional *supervision* be required?
  Will additional construction *equipment* be required?
  Will additional *materials* be required?
  Will additional *tools* be required?
  Will additional *material handling* be required?
  Will there be less *space* to work in?
  Will *other crews* create *congestion*?
  Has Rate Of Productivity (ROP) been affected? It can be recorded in list form, using the following headings:

Activity   ROP Before Change   ROP After Change

- Will additional clean-up be required?
- What *other factors* may affect productivity?
- Has a separate cost code been established?
- Has the crew leader been notified to maintain separate costs on a new cost code?

Having analyzed the actual time of the new work itself and the effect on time of how the work was performed, we then analyze time, considering the following points:

- What event on the CPM schedule is affected?
- Is installed equipment delayed?
- Was the crew *stopped* (or idle) in the field while waiting for direction? Which crews (or crew members) and for how long?
- Did the crew have to *demobilize* and *move* to another location?
- Did the crew have to *integrate* into another crew? Which crew and what were the effects? How long were the effects felt?
- Was the crew divided? What was the effect and for how long?
- Will this work take *longer* to perform? How much longer?
  Has hoist time availability been affected?
- Will this work have to be performed faster?
  Will *overtime* be required?
  Will *shift work* be required?
  Will additional manpower be required to make up the time?

Will additional construction be required to make up the time?

- Was construction equipment idle on standby?
- Will effort be required in a *later time period*, involving:
    Material escalation
    Labor escalation
    More difficult environment (winter work, etc.)
    Job site overhead costs that are time-related
    Is interest allowable?
    Has effect on home office overhead been considered?

Sequence

Part of this process is considering *how* the work is performed. In Chapter 6 on scheduling, a concept of production flow was discussed—sometimes called job "rhythm" or sequence. It is essential to analyze the construction schedule to determine if that sequence is being affected. Questions to consider are:

- Has crew movement (sequence) been altered? Which crew and how?
- Will material handling change?—lengthened or increased?
- Will productivity be affected because of change in crew sequence? Which crew and how?

Impact On Other Work (Unchanged Work)

In the previous steps, we considered the effect on only the changed work. Now we must consider the effect on the other crews—the unchanged work—by asking these questions:

- What other work activities will be *indirectly affected* by the change?
    - What crafts will be affected?
    - Which work area(s) will be affected?
    - For what duration?
- How will the other crafts or crews be affected?
    - Congestion of work space?
    - Interference with other crews?
    - Material handling?
    - Crew movements?
    - Demobilization?
    - Remobilization?
    - Crew size impact?
    - Learning curve?
    - Morale?
    - Overtime?
    - Shift work?
    - Re-work?
    - Clean-up?
    - Duration?
    - Supervision increase or dilution?
    - Is critical path affected?
    - Is float absorbed?
    - Will work sequence of other crafts be affected?

In summary, Figures 8.11 and 8.12 illustrate the effects that changes can have on a project—both on the changed work itself and on the unchanged work.

How to Prove Additional Costs

The owner or his representative are often distrustful of a contractor's cost proposal for a change impact, especially if it is not adequately supported by the contractor. The following are tools for helping the contractor identify and prove its costs.

To begin with, the contractor has the *burden of proving* its own cost or claim and must do so with adequate *documentation*. Adequate documentation means:

- *Updating the schedule* to show effect of the change.
- Establishing *separate cost codes* for changed or impacted work.
- Maintaining a system of *time sheets* reflecting the labor and equipment associated with the change.
- Maintaining *daily diaries* and other field records that show the field impact of a change.

Figure 8.13 is a useful tool for analyzing the schedule effect of a change. On this form, the contractor can show both the time effect, if any, on procurement, and the manpower input on the schedule by analyzing crew days. With this information, the contractor can now update his CPM or graph schedule.

Next is the Occurrence Report Form, shown in Figures 8.14a and 8.14b. This form can be used by the field supervisor to document

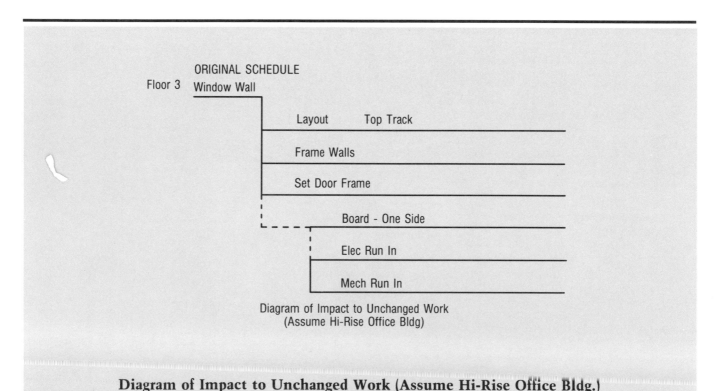

Diagram of Impact to Unchanged Work (Assume Hi-Rise Office Bldg.)
Window wall is modified, causing delay to window wall delivery and installation. General contractor directs drywaller subs to work in the core area and then come back down to pick up the perimeter of the building. See Figure 8.12 for an illustration of the effects of the change.

Figure 8.11

240

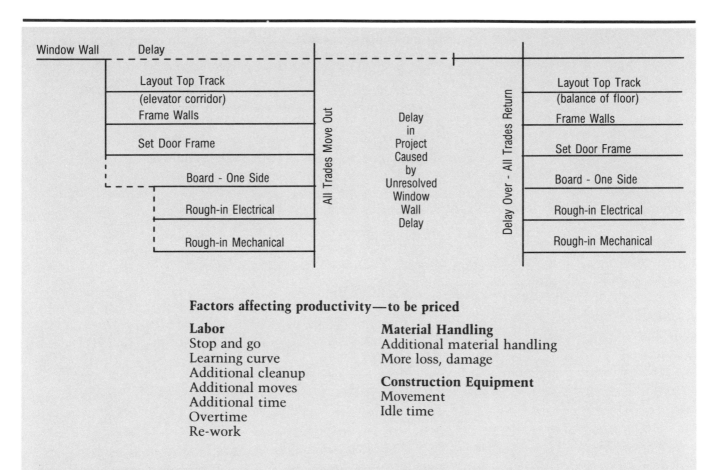

Factors affecting productivity—to be priced

Labor
Stop and go
Learning curve
Additional cleanup
Additional moves
Additional time
Overtime
Re-work

Material Handling
Additional material handling
More loss, damage

Construction Equipment
Movement
Idle time

The *changed work* was the window wall. The *unchanged work* is all the *craft crews* which must completely resequence their work. Remember, in the above diagram, this scenario is being *repeated* on each floor. The *same amount* of work is to be done, but it is being done in a *different manner*, for crews will be split, as it goes up and down the building. Material handling will be doubled.

Example of How a Change or Delay Can Also Affect the Unchanged Work

Figure 8.12

Change Proposal Review

	SUBMITTED	AMOUNT	RETURN REQUEST	APPROVAL STATUS — DATE	
#72	3-13-90	$30,258.00	3-27-90	App'd 4-10-90	AA - 293
#72m	9-26-89	$ 3,359.00	10-10-89	App'd 10-29-89	AA #304

DESCRIPTION: Adden. 177, 7-25-89, issued for pricing + construction
AD-10, AD-11, mo-02, PCA-287-291. Locates 5 beach showers.

CRAFTS AFFECTED:

Site Work - 214 MH w/5 men = 5^4 days   Paint   —

Concrete - 15 MH w/5 men = 0^4 days   Stone Paving - 540 MH w/5 men = 13^5 days

Carpentry - 50 MH w/4 men = 1^b days   Plumbing - 84 MH w/4 men = 2^6 days

                                             Total = 23^5 days

(old) Old Activity	(new) New Activity	(es) Early Start	(p) Preceeding Activity	(np) Notice to Proceed	(t) Duration	(cs) Critical Start	(s) Soft-Ware	(dt) Time Change	Actual	Total Days	
26	72	9-7-89	64	7-25-89		8-13-89	18	—	23		

SUBMIT CHANGE	BUY MATERIAL	DELIVER
3	3	12

Figure 8.13

each day what is actually happening on the job as a result of the change. A copy can be sent to the owner or his representative, thereby meeting the contractual requirements of notice.

To support their impact claims, contractors may refer to studies which show the effect outside influences (in the form of crew stacking, overtime, etc.) have on productivity. Figure 8.15 is taken from a publication of the Department of the Navy. This report presents a method for quantifying these outside impacts on the schedule.

Equipment and Tools

It is important that the contractor have an established system for recording the costs of owned construction equipment and small tools. It is recommended that the contractor use the AGC (Associated General Contractors' of America) *Contractors' Equipment Cost Guide*, the AED (Associated Equipment Distributors) *Rental Rates and Specifications*, or the *Rental Rate Blue Book for Construction Equipment*, published by Dataquest (Dunn & Bradstreet).

Unabsorbed Overhead

In cases of suspension of work on a project, it may be that the contractor's home office overhead is not being absorbed at the same rate it would have been had the compensable delay not occurred. The Eichleay Formula is widely used for this purpose, although it is now under some attack by the General Services Administration and others. The Eichleay Formula, based on a case entitled, "Eichleay Corp., ASBCA No. 5183,60-2 BCA II 2688,1960", is set forth below:

(1) $$\frac{\text{Contact Billings}}{\text{Total Billings}} \text{ (Actual Contract Period)} \times \text{Total Overhead for Contract Period} = \text{Overhead Allocable to Contract}$$

(2) $$\frac{\text{Allocable Overhead}}{\text{Days of Performance}} = \text{Daily Contract Overhead}$$

(3) $$\text{Daily Contract Overhead} \times \text{Number of Days Delay} = \text{Total Unabsorbed Home Office Overhead}$$

In addition, the contractor must show that it has been affected by the suspension of work, in terms of lost or diminished bonding or other effects on its ability to replace the revenue not earned during the suspension.

Clearly, there are other areas which are affected by changes, and other measures that should be taken to recover full costs. However, as this book is not intended to be a full treatise on construction claims, only the key areas have been covered. The form in Figure 8.16 lists all the items that should be addressed in pricing a change for a claim.

Occurrence Report

Date: _____   Job No. _____

Location of Occurrence: _____

_____

OCCURRENCE:

Change Order _____   Third Party _____
                                       (Owner, Owner's Rep., etc.)

Clarification _____   Subcontractor/Supplier _____

Design Deficiency _____   Weather _____

Shop Drawing _____   In House _____

Description of Occurrence: _____

_____

_____

_____

Impact of Occurrence (CPM activity affected): _____

_____

_____

_____

Field Action Required: _____

_____

_____

Figure 8.14a

Impact Checklist

Item	Yes	No	Comments
A. Crew			
1. Added Manpower			
2. Overtime			
3. Learning Curve Affected			
4. Crew Stacking			
5. Morale			
6. Movement (Demob./(Remob.)			
7. Stand-by			
8. Sequence Change			
B. Supervisor			
1. Added			
2. Insufficient Ratio			
C. Equipment			
1. Movement			
2. Stand-by			
3. Change of Method of Operation			

D. Production Rates

Cost Code Number and Work Activity	Budgeted Production Rate	Actual Production Rate

Figure 8.14b

Manpower

a. The two major impacts upon manpower are reduced productivity and pay scale increases. The latter is a factor when modifications delay progress to the point that work which would have otherwise been completed must now be performed at a time when higher wages are in effect. Reduced productivity takes many forms, and is therefore more difficult to quantify before the fact.

b. Reduced labor productivity implies a loss from some established normal or anticipated level of productivity. Although construction does not lend itself to definitive measurement of labor productivity, there are methods a contractor can use to quantify anticipated labor costs when preparing a bid. The most common technique draws heavily on data derived from the contractor's past experiences, including any indicated trends, present labor pay rates, and anticipated labor rate increases during the life of the project.

c. The contractor's NAS progress schedule carries lump sum values for each construction activity. However, the ratio of labor costs vs. material and equipment costs varies widely for different activities, so it is impractical to apply a universal rule-of-thumb ratio. Nevertheless, a fairly accurate breakout of labor costs for any activity can be obtained by subtracting from the total activity value all the cost items which are not direct labor. These items are material costs, equipment costs, and overhead and profit. The remaining costs are production labor costs, including wages paid, fringe benefits, insurance and taxes, and overhead and profit markups. Through a process of elimination, a reasonably accurate determination of the man-hours represented by the dollar amount of labor cost for an activity can be reached. The Government estimate for the project and data from the contractor's payroll submittals can also be helpful in projecting manpower levels on future activities.

d. That portion of the contract price devoted to labor costs indicates the contractor's anticipated level of labor productivity. Whether or not the anticipated profit can be realized from the completed project depends to a great extent on the contractor's ability to maintain the planned labor productivity level. With expert management and some good luck, the contractor may achieve labor productivity exceeding original expectations. Conversely, labor productivity or effectiveness can fall below expectations as a result of many uncontrollable factors.

e. The actual labor productivity of a project affects the cost of labor for modifications. On projects where actual labor productivity is running at or better than the contractor's anticipated level, data developed from the analysis described in c above is appropriate for pre-pricing direct and impact costs of modifications. However, when the contractor's actual labor costs are higher (productivity lower) than those anticipated by the bid, actual experience data should be considered. Depending on the degree to which contractor mismanagement has contributed to the higher labor costs, the estimator may find it expedient to use a combination of actual and anticipated productivity projections in arriving at a reasonable labor cost figure for the modification. This does not imply that a modification should be priced to reimburse the contractor for excess labor costs incurred because of his inept management; however, it is possible to incur labor costs higher than those anticipated. The higher costs can occur through no fault of the contractor. The estimator must therefore take this into account when forward-pricing direct or impact-related modification labor costs.

f. This pamphlet deals specifically with the estimation of the costs arising from impact on the unchanged work. It also encourages the settlement of contract modifications before the work is accomplished.

g. Prepricing of impact costs arising from labor is the most difficult aspect of the estimating process for two reasons. First, the estimator must verify that these determinations are reasonable and well founded. Second, when negotiating it is necessary to convince the contractor that the determinations are reasonable. Most contractors (and many personnel within the Corps of Engineers) would prefer to leave the settlement of impact cost/time until after the modification work is performed. However, for the reasons stated earlier, such a procedure is not recommended. The preferred approach is to anticipate the costs before the fact, and to include them in the cost estimate. Figures 4-1a through 4-4 illustrate the effects of various situations on construction manpower efficiency. These figures are included as a source of general information and some estimators may find them helpful in supplementing other data generated in the development of modification cost estimates. However, the validity of the graphs has not been sufficiently tested to warrant their use in preference to established methodologies.

Figure 8.15

h. The lowest reasonable price for modifications is estimated by basing the direct and impact costs of labor upon the productivity level established in e above. (Allowances for labor impact costs compensate the contractor for losses in productivity.) Typical causes of labor productivity loss on the unchanged work resulting from modifications are as follows.

(1) **Disruption**. The contractor's progress schedule represents the planned sequence of activities leading to final completion of the project. Workers who know what they are doing, what they will be doing next, and how their activities relate to the successful completion of the project develop a "job rhythm." Labor productivity is at its optimum when there is good job rhythm. When job rhythm is interrupted (i.e., when a contract modification necessitates a revision of the progress schedule), it affects workers on both the directly changed and/or unchanged work and may result in a loss of productivity.

(a) Disruption occurs when workers are prematurely moved from one assigned task to another. Regardless of the competency of the workers involved, some loss in productivity is inevitable during a period of orientation to a new assignment. This loss is repeated if workers are later returned to their original job assignment. Learning curves which graph the relationship between production rate and repeated performance of the same task have been developed for various industrial tasks. The basic principle of all learning curve studies is that efficiency increases as an individual or team repeats an operation over and over; assembly lines are excellent demonstrations of this principle. However, although construction work involves the repetition of similar or related tasks, these tasks are seldom identical. Skilled construction workers are trained to perform a wide variety of tasks related to their specific trade. Therefore, in construction it is more appropriate to consider the time required to become oriented to the task rather than acquiring the skill necessary to perform it. One of the attributes of the construction worker is the ability to perform the duties of his trade in a variety of environments. How long it will take the worker to adjust to a new task and environment depends on how closely related the task is to his experience or how typical it is to work usually performed by his craft. Figure 4-1a assumes that the worker will always be assigned to perform work within the scope of his trade, and that the average worker will require a maximum of one shift (8 hours) to reach full productivity. Full productivity (100 on the Theoretical Productivity Scale) represents optimum productivity for a given project. Figure 4-1b is a tabulation of productivity losses derived from figure 4-1a.

(b) The time required for a worker (or crew) to reach full productivity in a new assignment is not constant. It will vary with skill, experience, and the difference between the old and new task. In using the chart or its tabulation, the estimator must decide what point on the Theoretical Productivity Scale represents a composite of these factors. For example, an ironworker is moved from placing reinforcing bars to the structural steel erection crew. The ironworker is qualified by past training to work on structural steel, but the vast majority of his experience has been with rebars, and the two tasks are significantly different. In view of this, a starting point of "0" is appropriate. The estimator can determine from the chart that a "0" starting point indicates the ironworker will need 8 hours to reach full productivity, with a resulting productivity loss of 4 hours. The Government's liability is then 4 hours times the hourly rate times markups. As a second example, assume the same ironworker is moved from placing reinforcing bars for Building A to placing reinforcing bars for Building B. The buildings are similar but not identical. A starting point of "90" is appropriate. The duration of only 0.8 hours is required to reach full productivity, and the productivity loss is 0.4 hours. The Government's liability would then be 0.4 hours times the hourly rate times markups.

(c) The contractor normally absorbs many orientation/learning cycles as his labor forces are moved from task to task in the performance of the work. Only those additional manpower moves, caused solely by a contract modification, represent labor disruption costs for which the contractor is entitled extra payment.

(2) **Crowding**. If a contractor's progress schedule is altered so that more activities must be accomplished concurrently, impact costs caused by crowding can result. Crowding occurs when more workers are placed in a given area than can function effectively. Crowding causes lowered productivity; it can be considered a form of acceleration because it requires the contractor either to accomplish a fixed amount of work within a shorter time frame, or to accomplish more work within a fixed time frame. Granting additional time for completion of the project can eliminate crowding. When the final completion date cannot be slipped, increased stacking of activities must be analyzed and quantified.

Figure 8.15 (continued)

Manpower

(a) Activity stacking does not necessarily result in crowding — when concurrent activities are performed in areas where working room is sufficient, crowding is not a factor. But, if the modification forces the contractor to schedule more activities concurrently in a limited working space, crowding does result. Both increased activity stacking and limited (congested) working space must be present for crowding to become an item of impact cost.

(b) Crowding can be quantified by using techniques similar to those used for acceleration. Figure 4-2 illustrates the curve developed to represent increases in labor costs from crowding. Before applying this curve, the estimator must determine whether crowding will occur and to what degree. For example, the assumption that the contractor's scheduling of the activities in question is the most efficient sequencing of the work must be verified. Perhaps more workers can work effectively in the applicable work space than the contractor has scheduled; if they cannot, perhaps the crowding is not severe enough to justify using the full percentage of loss indicated by the graph. (The graph should be interpreted as representing the upper limit of productivity loss.) In this case, the estimator's judgment of the specific circumstances may indicate that some lower increase factor is appropriate.

(c) For example, assume that the estimator decides that severe crowding will occur in the following situation: The contractor's schedule indicates three activities concurrently in progress in a limited area of the project. Each of these activities employs five workers, placing a total of 15 workers in the area. One of these activities has a duration of 10 days; the other two have 20-day durations. The modification has required that a fourth activity be scheduled concurrently in the same limited area. This additional activity requires three workers; it has a normal duration of 5 days. There are now 18 workers in an area which can only efficiently accommodate 15. The percent of crowding is 3/15 or 20 percent. On the graph (figure 4-2), 20 percent crowding intersects the curve opposite 8 percent loss of efficiency. To find the duration of crowding, the estimator multiplies the normal duration of the added activity by 100 percent plus the percent loss of efficiency. For this example, 5 days times 1.08 equals 5.4 days. Therefore, because of the inefficiency introduced by crowding, the added activity will require 5.4 days to complete. Likewise, on the three affected activities, the first 5 days of normal activity will now require 5.4 days. All four activities will experience loss of productivity resulting from an inefficiency factor equivalent to 0.4 of a single day's labor cost. This is calculated as follows:

Average hourly rate x hours worked per day x number of workers x 0.4 = $ loss

or

$12.00 x 8 x 18 x 0.4 = $691 plus normal labor markups.

3/18 x $691 = direct crowding cost; and should be included in the Direct Cost section of the modification estimate

15/18 x $691 = crowding on unchanged activities, and should be placed in the Impact on Unchanged Work section of the modification estimate.

(3) **Acceleration**. Acceleration occurs when a modification requires the contractor to accomplish a greater amount of work during the same time period even though he may be entitled to an extension of time to accomplish the changed work. This is sometimes referred to as "buying back time." Acceleration should be distinguished from expediting. Expediting occurs whenever the modification would require the contractor to complete the work before the original completion date included in the contract. Per DAR 18-111, expediting is not permissible in the absence of approval by the Assistant Secretary of Defense (Manpower, Reserve Affairs, and Logistics). Acceleration may be accomplished in any of the following ways:

(a) *Increasing the size of crews*. The optimum crew size (for any construction operation) is the minimum number of workers required to perform the task within the allocated time frame. Optimum crew size for a project or activity represents a balance between an acceptable rate of progress and the maximum return from the labor dollars invested. Increasing crew size above optimum can usually produce a higher rate of progress, but at a higher unit cost. As more workers are added to the optimum crew, each new worker will increase crew productivity less than the previously added worker. Carried to the extreme, adding more workers will contribute nothing to overall crew productivity. Figures 4-3a through 4-3d indicate the effect of crew overloading.

Figure 8.15 (continued)

248

(b) ***Increasing shift length and/or days worked per week***. The standard work week is 8 hours per day, 5 days per week (Monday through Friday). Working more hours per day or more days per week introduces premium pay rates and efficiency losses. Workers tend to pace themselves for longer shifts and more days per week. An individual or a crew working 10 hours a day, 5 days a week, will not produce 25 percent more than they would working 8 hours a day, 5 days a week. Longer shifts will produce some gain in production, but it will be at a higher unit cost than normal hour work. When modifications make it necessary for the contractor to resort to overtime work, some of the labor costs produce no return because of inefficiency. Costs incurred due to loss of efficiency created by overtime work are an impact element because the increase in overtime results from the introduction of the modification. Contractors occasionally find that to attract sufficient manpower and skilled craftsmen to the job, it is necessary to offer overtime work as an incentive. When this is done, the cost must be borne by the contractor; however, if overtime is necessary to accomplish modification work, the Government must recognize its liability for introducing efficiency losses. Figure 4-4 is the result of a study which attempted to graphically demonstrate efficiency losses over a 4-week period for several combinations of work schedules. These data are included merely as information on trends rather than firm rules which might apply to any project. Although figure 4-4 data do not extend beyond the fourth week, it is assumed that the curves would flatten to a constant efficiency level as each work schedule is continued for longer periods of time.

(c) ***Multiple shifts***. The inefficiencies in labor productivity caused by overtime work can be avoided by working two or three 8-hour shifts per day. However, additional shifts introduce other costs. These costs would include additional administrative personnel, supervision, quality control, lighting, etc. Modifications that cause the contractor to implement shift work should price the impact cost as appropriate for the activity being accelerated. Environmental conditions such as lighting and cold weather may also influence labor efficiency.

(4) **Morale**. The responsibility for motivating the work force and providing a psychological environment conducive to optimum productivity rests with the contractor. Morale does exert an influence on productivity, but so many factors interact on morale that their individual effects defy quantification. A project's contract modifications, particularly a large number, have an adverse effect on the morale of the workers. The degree to which this may affect productivity, and consequently the cost of performing the work, would normally be very minor when compared to the other causes of productivity loss. A contractor would probably find that it would cost more to maintain the records necessary to document productivity losses from lowered morale than justified by the amount he might recover. Modification estimates do not consider morale as a factor because whether morale becomes a factor is determined by how effective the contractor is in his labor relations responsibilities.

Quantification. The following example demonstrates how to use figures 4-3a through 4-3d to quantify the impact costs of crew overloading. Assume that the contractor has planned a construction operation with a duration of 15 working days and an optimum crew size of 10. The modification now requires that the contractor accomplish this operation in 10 working days. The rate of production is the unit of work per amount of effort in man days. The percent increase is new rate minus original rate divided by original rate times 100. Thus,

$$\frac{(1 \text{ job} \div 100 \text{ MD}) - (1 \text{ job} \div 150 \text{ MD})}{1 \text{ job} \div 150 \text{ MD}} \times 100 = \frac{.01 - .0067}{.0067} \times 100 = 50 \text{ percent}$$

Figure 8.15 (continued)

Manpower

This represents a 50 percent increase in the crew's rate of production. From figure 4-3a or 4-3d, it appears likely that 50 percent production gain can be achieved by increasing the size 80 percent. Other options could be implemented to speed up production: the optimum crew could work longer shifts, more days per week; a second crew could be placed in operation (if allowed by the nature of the work). However, for this example only increasing crew size is considered. The way to quantify the impact cost before the fact is:

	Original Plan	Accelerated Plan
Manpower	10	18
Hourly Rate	$12	$12
Crew Cost/Day (8 hours)	$960	$1,728
Duration (Working Days)	15	10
Crew Cost (Cost/Day x Duration)	$14,400	$17,280
Taxes, Insurance, Fringes (18 percent)	$ 2,595	$ 3,110
Total Crew Cost	$16,992	$20,390

Impact Cost (Accelerated-Original) = $3,398 ($3,400)

      — or —

Impact Cost (Accelerated Plan x Efficiency Loss) =
      $20,390 x 16.7 percent (from fig. 4-3b) = $3,405 ($3,400)

The amount of $3,400 would be placed in the modification estimate, under "Impact on Unchanged Work" and identified by the activity involved. Increased cost of supervision, if necessary, is not included in this crew overloading analysis. Supervision must be costed separately, either as a separate item or as an element of Job Site Overhead, as appropriate.

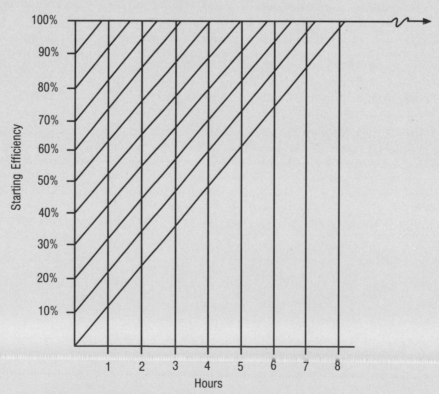

Note: 100% represents the productivity rate required to maintain scheduled progress.

Construction Operations Orientation/Learning Chart

Figure 8.15 (continued)

250

Productivity Losses Derived from Construction Operations Orientation/Learning Chart

Productivity Starting Point	Duration (Hr.)	Average Loss (Hr.)
100	0	0
90	0.8	0.4
80	1.6	0.8
70	2.4	1.2
60	3.2	1.6
50	4.0	2.0
40	4.8	2.4
30	5.6	2.8
20	6.4	3.2
10	7.2	3.6
0	8.0	4.0

Productivity Losses

Figure 8.15 (continued)

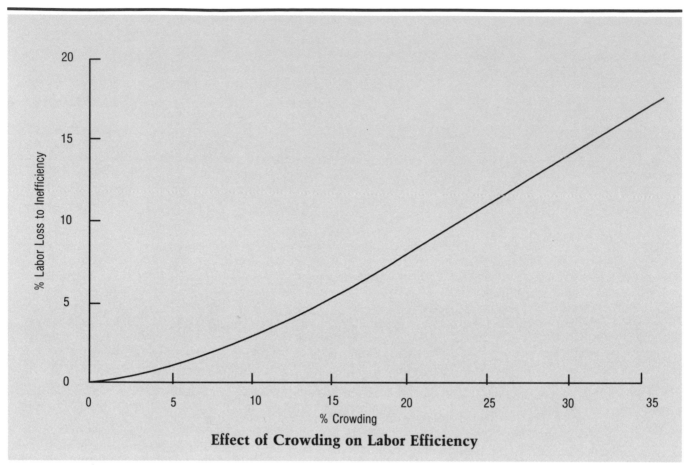

Effect of Crowding on Labor Efficiency

Figure 8.15 (continued)

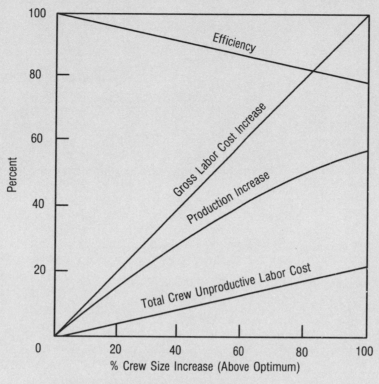

Composite Effects of Crew Overloading

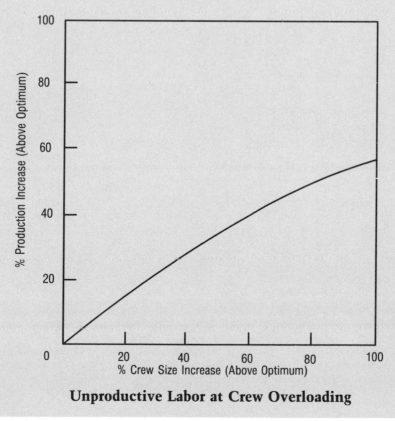

Unproductive Labor at Crew Overloading

Figure 8.15 (continued)

253

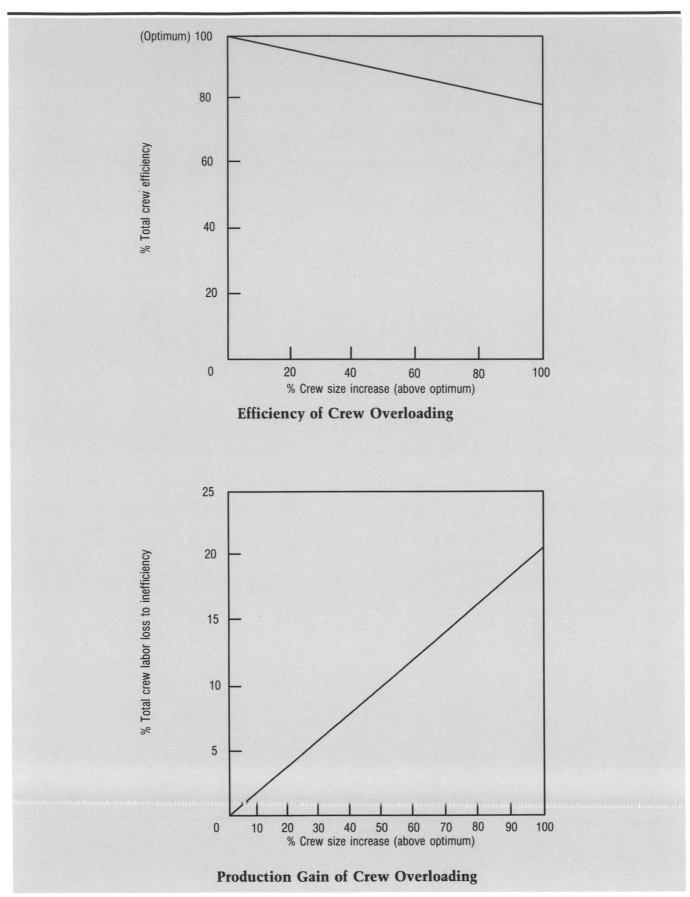

Efficiency of Crew Overloading

Production Gain of Crew Overloading

Figure 8.15 (continued)

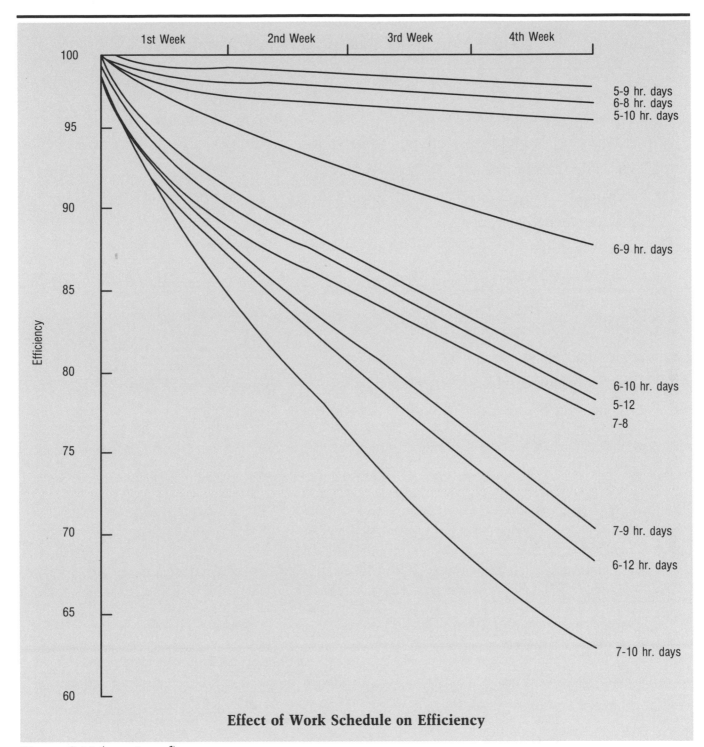

Effect of Work Schedule on Efficiency

Figure 8.15 (continued)

The Elements of Pricing Changed Work

Labor

☐ Additional labor for *added tasks*

☐ Additional labor for longer duration to do work

☐ Delay due to stop and go

☐ Rework

☐ Loss of productivity due to:

    ___ Overtime

    ___ Winter work

    ___ Crew congestion

    ___ Overbalanced crew

    ___ Split crews

    ___ Learning curve

☐ Application of wage rates including escalation

☐ Taxes and insurance

☐ Union fringes

☐ Supervision

☐ Subsistence

☐ Travel expense

Note: Document with Time Sheets, Occurrence Reports, Updated Schedule

Materials

☐ Scrap existing material

☐ Additional material

☐ Material escalation

☐ Material storage

☐ Warranty extension cost

☐ Shipping charges

☐ Taxes

Note: Document with invoices, receipts.

Figure 8.16

The Elements of Pricing Changed Work

Construction Equipment

☐ Stand-by

☐ Additional time

☐ Productivity impact due to:

    ____ More confined workspace

    ____ Shorter (or longer) haul distances

    ____ Weather conditions

☐ Rentals—leases and receipts

Note: Document with Equipment Time Sheets, Occurrence Reports, *AGC Contractor's Equipment Cost Guide,* or other standard references.

Job Site Overhead

☐ Identify delay period

☐ Develop daily rate for job site overhead expenses

Interest (where allowed)

☐ Show interest on borrowed amount for *this* project

Home Office Overhead

☐ On government jobs, use audited rate audited by OCAA

☐ On non-government jobs, use CPA developed rate

☐ Use Eichleay Formula for unabsorbed overhead

Bond Premium

Insurance Premium

Mark-up on Subcontractor's Cost

Subcontractor's Cost

Profit

Figure 8.16 (continued)

Summary

Changes are inevitable in the construction industry. Therefore, change order and claim management are essential if a contractor is to protect his profit. Owners often resist paying the cost of changes, and contractors seldom do an adequate job in proving what they are worth. The owner may thereby have an advantage, knowing the average contractor will not do an adequate job of contract notification, or field documentation. Indeed, it is the *average* contractor who becomes a part of the enormous turnover of contractors who can build but do not know how to manage.

Appendix

Appendix

The following thirty pages are the Means Spec-Aid form. Spec-Aid serves a variety of purposes: it is a preliminary checklist for detailed project specifications; it is a project guideline which allows the user to clearly define and analyze a project's scope and details; and, as referenced in Chapter 3, it is a project checklist to confirm that all items are included in the estimate/bid as called for in the project plans and specifications.

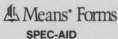

SPEC-AID

PROJECT _____ DATE _____

ADDRESS _____

CITY/STATE/ZIP _____ PREPARED BY: _____

PROJECT SKETCH 1" = _____ Ft.

Means Forms

SPEC-AID

DIVISION 1: GENERAL

PROJECT _____ LOCATION _____

Owner _____ **Architect** _____ **Project Mgr.** _____

Engineer: Structural _____ Plumbing _____

   H.V.A.C. _____ Electrical _____

Contractor: General _____ Structural _____

   Mechanical _____ Electrical _____

Building Type _____

Building Capacities: _____

Quality ☐ Economy ☐ Average ☐ Good ☐ Luxury   Describe _____

Size

  Ground Floor Area _____ S.F.

  Supported Levels (No.) _____ x Area/Level _____ S.F.

  Supported Levels (No.) _____ x Area/Level _____ S.F.

  Below Grade Area _____ S.F.

  Other Area _____ S.F.

                    TOTAL GROSS AREA _____ S.F.

  Floor to Floor Height: Maximum _____ Minimum _____ Average _____

  Floor to Ceiling Height: Maximum _____ Minimum _____ Average _____

  Floor System Depth: Maximum _____ Minimum _____ Average _____

Building Codes ☐ City _____ ☐ County _____

  ☐ State _____ ☐ National _____

  ☐ Other _____ Seismic Zone _____

Zoning ☐ Residential ☐ Commercial ☐ Industrial ☐ None ☐ Other _____

Design Criteria Live Loads: Roof _____ psf. End Walls _____ psf. Window Openings _____ %

  Supported Floor _____ psf. Side Walls _____ psf. Window Openings _____ %

  Ground Floor _____ psf.   WIND PRESSURE

  Corridors _____ psf. _____ psf. from _____ to _____ ft.

  Balconies _____ psf. _____ psf. from _____ to _____ ft.

  Allow for Partitions _____ psf. _____ psf. from _____ to _____ ft.

  Miscellaneous _____ psf. _____ psf. from _____ to _____ ft.

Comments _____

Typical Bay Spacing _____

Structural Frame ☐ Concrete ☐ Steel ☐ Wood ☐ Wall Bearing ☐ Other _____

  Describe _____

Fireproofing ☐ None ☐ Columns _____ Hours ☐ Girders _____ Hours ☐ Beams _____ Hours ☐ Floor _____ Hours

Estimating Budget Estimate Due _____ 19 _____ Schematic Estimate Due _____ 19 _____

  Preliminary Estimate Due _____ 19 _____ Final Estimate Due _____ 19 _____ at _____ % Working Drawings

Labor Market ☐ Highly Competitive ☐ Normal ☐ Non-Competitive ☐ Unreliable ☐ Union ☐ Non-union

  Describe _____

Taxes Tax exempt ☐ No ☐ Yes State ____% County ____% City ____% Other ____%

Bond ☐ Not Required ☐ Required _____

Bidding Date _____ Start Date _____ Construction Duration _____ Months

  ☐ Open Competitive ☐ Selected Competitive ☐ Negotiated ☐ Filed Bids _____

Contract ☐ Single ☐ Multiple Describe _____

  Multiple Type assigned to General Contractor ☐ No ☐ Yes _____

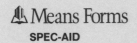 **Means Forms**

SPEC-AID

DATE _____

DIVISION 2: SITEWORK

PROJECT _____ LOCATION _____

Demolition Site: ☐ No ☐ Yes Allowance _____ ☐ Separate Contract
    Interior: ☐ No ☐ Yes ☐ Allowance _____ ☐ Separate Contract
    Removal From Site: ☐ No ☐ Yes Dump Location _____ Distance _____

Topography ☐ Level ☐ Moderate Grades ☐ Steep Grades Describe _____

Subsurface Exploration ☐ Borings ☐ Test Pits ☐ USDA Maps ☐ Other _____
    Performed by: ☐ Owner ☐ Engineer ☐ Contractor _____

Site Area: Total _____ Acres to Clear _____ Acres To Thin _____ Acres Open _____ Acres

Clearing and Grubbing: ☐ No ☐ Light ☐ Medium ☐ Heavy _____

Topsoil: ☐ No ☐ Strip ☐ Stockpile ☐ Dispose on Site ☐ Dispose off Site _____ Miles ☐ Furnish
    Existing _____ Inches Deep Final Depth _____ Inches Describe _____

_____

Soil Type: ☐ Gravel ☐ Sand ☐ Clay ☐ Silt ☐ Rock ☐ Peat ☐ Other _____
    Rock Expected: ☐ No ☐ Ledge ☐ Boulders ☐ Hardpan ☐ Describe _____
    How Paid _____
    Ground Water Expected: ☐ No ☐ Yes Depth or Elevation _____
    Disposal by ☐ Pumping ☐ Wells ☐ Wellpoints ☐ Other _____

Excavation: ☐ Grade and Fill on Site ☐ Dispose off Site _____ Miles ☐ Borrow Expected _____ Miles
    Quantity Involved _____
    Describe _____
    Sheeting Required: ☐ No ☐ Yes Describe _____
    Protect Existing Structures: ☐ No ☐ Yes Describe _____

Backfill: ☐ No ☐ Yes Area _____ Material _____ Inches Deep _____ % Compaction
    Landscape Area ☐ No ☐ Yes Area _____ Material _____ Inches Deep _____ % Compaction
    Building Area ☐ No ☐ Yes Area _____ Material _____ Inches Deep _____ % Compaction
    Source of Materials _____

Water Control: ☐ Ditching ☐ Sheet Piling ☐ Pumping ☐ Wells ☐ Wellpoints ☐ Pressure Grouting
    ☐ Chemical Grouting ☐ Other _____
    Describe _____

Termite Control: ☐ No ☐ Yes Describe _____

Special Considerations: _____

_____

Piles: ☐ No ☐ Yes ☐ Friction ☐ End Bearing ☐ Concrete ☐ Pipe, Empty ☐ Pipe, Concrete Filled ☐ Steel
    ☐ Step Tapered ☐ Tapered Thin Shell ☐ Wood ☐ Capacity _____ Tons
    Size _____ Length _____ Number Required _____

Caissons: ☐ No ☐ Yes ☐ Cased ☐ Uncased Capacity _____
    Size _____ Length _____ Number Required _____

Pressure Injected Footings: ☐ No ☐ Yes ☐ Cased ☐ Uncased Capacity _____
    Size _____ Length _____ Number Required _____

Special Considerations: _____

_____

Storm Drains: ☐ No ☐ Yes ☐ Asbestos Cement ☐ Bituminous Fiber ☐ Concrete ☐ Corrugated Metal ☐ _____
    Size and Length _____
    Headwall: ☐ No ☐ Yes Type _____ Number _____
    Catch Basins: ☐ No ☐ Yes ☐ Block ☐ Brick ☐ Concrete ☐ Precast Size _____ Number _____
    Manholes: ☐ No ☐ Yes ☐ Block ☐ Brick ☐ Concrete ☐ Precast Size _____ Number _____

Building Sub Drains: ☐ No ☐ Yes ☐ Type _____ Length _____

French Drains: ☐ No ☐ Yes Size _____ Length _____

Trenches: Swales, etc.: ☐ No ☐ Yes Describe _____

Rip Rap: ☐ No ☐ Yes Describe _____

Special Considerations: _____

_____

Water Supply Existing Main: ☐ No ☐ Yes Location _____ Size _____
  Service Piping: By Utility ☐ By Others ☐ This Contract _____ Size _____
  Wells: ☐ No ☐ Yes ☐ By Others ☐ This Contract _____ Capacity _____
  Water Pumping Station: ☐ No ☐ Yes Type _____ Capacity _____
Sewers: ☐ No ☐ Yes ☐ By Others ☐ Asbestos Cement ☐ Concrete ☐ Plastic ☐ Vitrified Clay ☐ _____
  Manholes: ☐ No ☐ Yes ☐ Block ☐ Brick ☐ Concrete ☐ Precast Size _____ Number _____
  Sewage Pumping Station: ☐ No ☐ Yes Type _____ Capacity _____
Sewage Treatment: ☐ No ☐ Yes ☐ By Others ☐ This Contract ☐ Septic Tank ☐ Package Treatment Plant
  Describe _____
Special Considerations: _____
_____

Driveways: ☐ No ☐ Yes ☐ By Others ☐ Bituminous ☐ Concrete ☐ Gravel ☐ _____ Thickness _____
Parking Area: ☐ No ☐ Yes ☐ By Others ☐ Bituminous ☐ Concrete ☐ Gravel ☐ _____ Thickness _____
  Base Course: ☐ No ☐ Yes ☐ By Others ☐ Gravel ☐ Stone ☐ _____ Thickness _____
Curbs: ☐ No ☐ Yes ☐ By Others ☐ Bituminous ☐ Concrete ☐ Granite ☐ _____ Size _____
Parking Bumpers: ☐ No ☐ Yes ☐ By Others ☐ Concrete ☐ Timber _____
Painting Lines: ☐ No ☐ Yes ☐ By Others ☐ Paint ☐ Thermo Plastic ☐ Traffic Lines ☐ Stalls ☐ _____
Guard Rail: ☐ No ☐ Yes ☐ By Others ☐ Cable ☐ Steel ☐ Timber _____
Sidewalks: ☐ No ☐ Yes ☐ By Others ☐ Bituminous ☐ Brick ☐ Concrete ☐ Stone _____
  Width _____ Thickness _____
Steps: ☐ No ☐ Yes ☐ Brick ☐ Concrete ☐ Stone ☐ Timber _____
Signs: ☐ No ☐ Yes ☐ Stock ☐ Custom _____
Traffic Signals: ☐ No ☐ Yes ☐ By Others _____
Special Considerations: _____
_____

Fencing: ☐ No ☐ Yes ☐ By Others ☐ Chain link ☐ Aluminum ☐ Steel ☐ Other _____
  Height _____ Length _____ Gates _____
Fountains: ☐ No ☐ Yes ☐ By Others _____
Planters: ☐ No ☐ Yes ☐ By Others ☐ Asbestos Cement ☐ Concrete ☐ Fiberglass ☐ _____
Playground Equipment: ☐ No ☐ Yes ☐ By Others ☐ Benches _____ ☐ Bleachers _____
  ☐ Bike Rack _____ ☐ Goal Posts _____ ☐ Posts _____
  ☐ Running Track _____
  ☐ See Saw _____ ☐ Shelters _____ ☐ Slides _____
  ☐ Swings _____ ☐ Whirlers _____ ☐ _____
Playing Fields: ☐ No ☐ Yes ☐ By Others _____
Railroad Work: ☐ No ☐ Yes ☐ By Others Weight _____ lb. per L.Y. ☐ New ☐ Relay Length _____
  Turnout: ☐ No ☐ Yes ☐ Bumpers _____ ☐ Derails _____
  Wheel Stops _____ ☐ Others _____
Retaining Walls: ☐ No ☐ Yes ☐ By Others ☐ Gravity Concrete ☐ Cantilever Concrete ☐ Steel Bin ☐ Cribbing
  ☐ Timber ☐ Other _____ Height _____ Length _____
Irrigation System: ☐ No ☐ Yes ☐ By Others _____
Tennis Courts: ☐ No ☐ Yes ☐ By Others Type _____ Number _____
Trash Closures: ☐ No ☐ Yes Size _____
Lawns & Planting: ☐ No ☐ Yes ☐ By Others _____ Allowance _____
  Topsoil: ☐ No ☐ Yes ☐ By Others Depth _____ Inches Source _____
  Shrubs: ☐ No ☐ Yes ☐ By Others Describe _____ Allowance _____
  Trees: ☐ No ☐ Yes ☐ By Others Describe _____ Allowance _____
  Seeding: ☐ No ☐ Yes ☐ By Others Describe _____
  Sodding: ☐ No ☐ Yes ☐ By Others Describe _____ Thickness _____
  ☐ Ground Cover _____ ☐ Edging _____ ☐ Mulching _____
Special Considerations: _____

DATE _____

DIVISION 3: CONCRETE

PROJECT _____ LOCATION _____

Foundations Bearing on: ☐ Rock ☐ Earth ☐ Piles ☐ Caissons ☐ Other _____
Bearing Capacity _____

Footings Pile Caps: ☐ No ☐ Yes _____ psi Size _____
    Forms _____ Reinforcing _____ Waterproofing _____
    Spread Footings: ☐ No ☐ Yes ___ psi Size _____ Soil Bearing Capacity _____
    Forms _____ Reinforcing _____ Waterproofing _____
    Continuous Footings: ☐ No ☐ Yes _____ psi Size _____
    Forms _____ Reinforcing _____ Waterproofing _____
    Grade Beams: ☐ No ☐ Yes _____ psi Size _____
    Forms _____ Reinforcing _____ Waterproofing _____

Piers: ☐ No ☐ Yes _____ psi Size _____
    Forms _____ Reinforcing _____ Finish _____

Anchor Bolts: ☐ No ☐ Yes Size _____
    Grout Column Base Plates: ☐ No ☐ Yes _____

Underslab Fill: ☐ No ☐ Yes Material _____ Depth _____
Vapor Barrier: ☐ No ☐ Yes Material _____ Thickness _____
Perimeter Insulation: ☐ No ☐ Yes Material _____ Dimensions _____
Slab on Grade: ☐ No ☐ Yes _____ psi Thickness _____
    Forms: ☐ Cold Keyed ☐ Expansion ☐ Other _____ Spacing _____
    Reinforcing: ☐ No ☐ Mesh ☐ Bars _____ Type _____
    Finish: ☐ Screed ☐ Darby ☐ Float ☐ Broom ☐ Trowel ☐ Granolithic _____
    Special Finish: ☐ No ☐ Hardner ☐ Colors ☐ Abrasives ☐ _____

Columns: ☐ No ☐ Round ☐ Square ☐ Rectangular ☐ Precast ☐ Steel ☐ Encased Steel ☐ Lightweight
_____ psi Size _____
    Forms: ☐ Optional ☐ Framed Plywood ☐ Plywood ☐ Fiber Tube ☐ Steel ☐ Round Fiberglass ☐ _____
    Reinforcing: ☐ No ☐ Square Tied ☐ Spirals Grade _____ Bar Sizes _____ Type Splice _____
    Finish: ☐ Break Fins ☐ Rubbed ☐ Other _____

Elevated Slab System: ☐ No ☐ Flat Plate ☐ Flat Slab ☐ Domes ☐ Pans ☐ Beam & Slab ☐ Lift Slab ☐ Composite
    ☐ Floor Fill ☐ Roof Fill ☐ Standard Weight ☐ Lightweight Concrete Strength _____
    Forms: ☐ Optional ☐ Plywood ☐ Other _____ Ceiling Height _____
    Reinforcing: ☐ Mesh ☐ Bars Grade _____ Size: _____
    Post-tension: ☐ No ☐ Simple Spans ☐ Continuous Spans _____ Depth _____
    ☐ Grouted ☐ Ungrouted Perimeter Conditions _____
    Slab Finish: ☐ Screed ☐ Darby ☐ Float ☐ Broom ☐ Trowel ☐ Granolithic ☐ _____
    Special Finish: ☐ No ☐ Hardener ☐ Colors ☐ Abrasives ☐ _____
    Ceiling Finish: ☐ No ☐ Break Fins ☐ Rubbed ☐ Other _____

Beams: ☐ No ☐ Steel ☐ Encased Steel ☐ Precast ☐ Regular Weight ☐ Lightweight ☐ Steel Composite _____
    Description: _____
    Forms: ☐ Optional ☐ Framed Plywood ☐ Plywood ☐ Steel ☐ Other _____ Ceiling Height _____
    Reinforcing: ☐ Conventional ☐ Post-tension ☐ Simple Span ☐ Continuous Spans _____ Depth _____
    ☐ Grouted ☐ Ungrouted Perimeter Conditions _____

Walls: ☐ No ☐ Precast ☐ Tilt up ☐ Regular Weight ☐ Lightweight _____ psi Thickness _____
    Forms: ☐ Optional ☐ Framed Plywood ☐ Plywood ☐ Steel ☐ Slipform ☐ _____
    Reinforcing: ☐ No ☐ Bars Grade _____ Clear Height _____
    Finish: ☐ No ☐ Break Fins ☐ Rubbed ☐ _____

Stairs: ☐ No ☐ Precast ☐ Ground Cast ☐ Form Cast ☐ Pan Fill Treads ☐ _____
    Forms: ☐ Plywood ☐ Steel ☐ Prefab Steel, Left in Place ☐ _____
    Reinforcing: ☐ Conventional Grade _____
    Finish: ☐ All Surfaces ☐ Treads ☐ Risers ☐ Abrasives ☐ Nosings _____

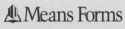 Means Forms

SPEC-AID

<u>**DIVISION 3: CONCRETE**</u>

Reinforcing Splices: ☐ No ☐ Yes ☐ Lap Type ☐ Compression Only ☐ 125% Yield ☐ Full Tension
☐ Horizontal ☐ Vertical ☐ Special _____

Gunite: ☐ No ☐ Yes

Cast in Place Special Considerations: _____
_____
_____
_____

Copings: ☐ No ☐ Yes Size _____ Finish _____

Curbs: ☐ No ☐ Yes Size _____ Finish _____

Joists: ☐ No ☐ Yes Live load _____psf. Span _____Size _____
Describe _____

Lift Slab: ☐ No ☐ Yes No. Slabs _____Thickness _____Inches Columns Spacing _____
Story Height_____ ☐ Conventional Reinforcing ☐ Post-tension _____

Lintels: ☐ No ☐ Yes ☐ Doors ☐ Windows ☐ Other _____
☐ Conventional Reinforcing ☐ Prestressed _____

Prestressed Precast Floors: ☐ No ☐ Yes ☐ Plank ☐ Multiple Tee Depth _____Span _____Width _____
Roofs: ☐ No ☐ Yes ☐ Plank ☐ Double Tee ☐ Single Tee Depth _____Span _____Width _____
Supporting Beams: ☐ No ☐ Yes ☐ Cast in Place ☐ Precast Describe _____

Columns: ☐ No ☐ Yes ☐ Steel ☐ Cast in Place ☐ Precast Size _____
Walls: ☐ No ☐ Yes ☐ Multiple Tee ☐ Other Thickness _____Height _____Width _____

Stairs: ☐ No ☐ Yes ☐ Treads Only ☐ Tread & Riser Units ☐ Complete Stairs ☐ Other _____
Describe _____

Tilt Up Walls: ☐ No ☐ Yes Size _____ Finish _____

Wall Panels: ☐ No ☐ Yes ☐ Insulated ☐ Regular Weight ☐ Lightweight Panel Size _____
Finish: ☐ Gray ☐ White ☐ Exposed Aggregate ☐ Other _____
Reinforcing: ☐ Conventional ☐ Prestressed ☐ Plain ☐ Galvanized _____
Erection: ☐ No. Stories _____Maximum Lift _____Overhangs, etc. _____

Window Section: ☐ No ☐ Yes ☐ Size _____ Finish _____

Window Sills: ☐ No ☐ Yes Size _____ Finish _____

Precast Special Considerations: _____
_____
_____
_____
_____
_____

Concrete Decks: ☐ No ☐ Yes ☐ Cast in Place ☐ Plank ☐ Topping ☐ Cement Fiber ☐ Channel Slab ☐ _____
Depth _____ Span _____ Sub Purlins _____ Roof Pitch _____

Concrete Fill: ☐ No ☐ Yes ☐ Regular Weight ☐ Lightweight Type _____ Depth _____

Formboard: ☐ No ☐ Yes ☐ Type _____ Depth _____ Spans _____
Sub Purlins: ☐ No ☐ Yes Span _____ Describe _____

Gypsum Decks Floor Plank: ☐ No ☐ Yes Depth _____ Span _____ Underlayment _____
Roofs: ☐ No ☐ Yes ☐ Cast in Place ☐ Plank Depth _____ Span _____ Pitch _____

Other Cementitious Decks: ☐ No ☐ Yes Describe _____

Cementitious Decks: Special Considerations _____
_____
_____

General Notes: Concrete Section _____
_____
_____

Means Forms

SPEC-AID

DATE _____

DIVISION 4: MASONRY

PROJECT _____ LOCATION _____

Exterior Walls: ☐ No ☐ Yes ☐ Load Bearing ☐ Non Load Bearing Story Height _____
Describe _____

Interior Walls: ☐ No ☐ Yes ☐ Load Bearing ☐ Non Load Bearing Ceiling Height _____
Describe _____

Mortar: ☐ Optional ☐ Type K ☐ Type 0 ☐ Type N ☐ Type S ☐ Type M ☐ Thinset ☐ _____
☐ Colors _____ ☐ Other _____

Cement Brick: ☐ No ☐ Yes ☐ Solid ☐ Cavity ☐ Veneer ☐ _____
Describe _____
_____ Compressive Strength _____ psi. ASTM No. _____
Size _____ Bond _____ Joints _____ Reinforcing _____ Ties _____

Common Brick: ☐ No ☐ Yes ☐ Solid ☐ Cavity ☐ Veneer ☐ _____
Describe _____
_____ Compressive Strength _____ psi. ASTM No. _____
Size _____ Bond _____ Joints _____ Reinforcing _____ Ties _____

Face Brick: ☐ No ☐ Yes ☐ Solid ☐ Cavity ☐ Veneer ☐ _____
Describe _____
_____ Compressive Strength _____ psi. ASTM No. _____
Size ☐ Standard ☐ Jumbo ☐ Norman ☐ Roman ☐ Engineer ☐ Double ☐ _____
Allowance: ☐ No ☐ Yes $ _____ per M Delivered ☐ Unglazed ☐ Single Glazed ☐ Double Glazed
Bond: ☐ Running ☐ Common ☐ English ☐ Flemish ☐ Stack ☐ _____ Headers Every _____ Course.
Joints: ☐ Concave ☐ Struck ☐ Flush ☐ Raked ☐ Weathered ☐ Stripped ☐ _____
Reinforcing: ☐ No ☐ Yes Describe _____
Wall Ties: ☐ No ☐ Yes Describe _____

Anchor Bolts: ☐ No ☐ Yes Size _____

Chimneys: ☐ No ☐ Yes ☐ Regular Brick ☐ Radial Brick Size _____

Columns: ☐ No ☐ Yes Size _____

Control Joints: ☐ No ☐ Yes Spacing _____ Material _____

Copings ☐ No ☐ Yes ☐ Concrete ☐ Stone ☐ _____ Describe _____

Fire Brick: ☐ No ☐ Yes ☐ Low Duty ☐ High Duty Describe _____

Fireplaces: ☐ No ☐ Yes Describe _____
Accessories _____

Flooring: ☐ No ☐ Yes ☐ Laid Flat ☐ Laid on Edge ☐ Pattern _____
☐ Regular ☐ Acid Resisting Describe _____

Insulating Brick: ☐ No ☐ Yes Describe _____

Insulation: ☐ No ☐ Yes ☐ Board ☐ Poured ☐ Sprayed Material _____
Thickness _____ Describe _____

Lintels: ☐ No ☐ Yes ☐ Block ☐ Precast ☐ Steel Describe _____

Masonry Restoration: ☐ No ☐ Yes ☐ Cut ☐ Recaulk ☐ Repoint ☐ Stucco Finish ☐ _____
Sand Blast: ☐ No ☐ Yes Describe _____
Steam Clean: ☐ No ☐ Yes Describe _____

Piers: ☐ No ☐ Yes Size _____

Pilasters: ☐ No ☐ Yes Size _____

Refractory Work: ☐ No ☐ Yes Describe _____

Simulated Brick: ☐ No ☐ Yes Material _____ Describe _____

Steps: ☐ No ☐ Yes Describe _____

Vent Box: ☐ No ☐ Yes ☐ Aluminum ☐ Bronze Size _____

Weep Holes: ☐ No ☐ Yes Spacing _____ Describe _____

Window Sills and Stools: ☐ No ☐ Yes ☐ Brick ☐ Concrete ☐ Stone ☐ _____
Describe _____

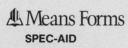 Means Forms

SPEC-AID

DIVISION 4: MASONRY

Concrete Block: □ No □ Yes □ Exterior □ Interior □ Regular Weight □ Lightweight □ Solid □ Hollow
□ Load Bearing □ Non Load Bearing Describe _____
_____ Compressive Strength _____ psi. ASTM No. _____
Size: _____
Finish: □ Regular □ Ground □ Ribbed □ Glazed □ _____
Bond: □ Common □ Stack □ Other _____ Headers Every _____ Course.
Joints: □ Concave □ Struck □ Flush □ Raked □ Weathered □ Stripped □ _____
Reinforcing: □ No □ Yes _____ Strips Every _____ Course.
Wall Ties: □ No □ Yes Describe _____
Bond Beams: □ No □ Yes Size _____ Describe _____
  Reinforcing _____ Grout _____
Lintels: □ No □ Yes □ Precast □ Steel □ Block Size _____ Describe _____
Reinforcing _____ Grout _____
Columns: □ No □ Yes Size _____
Pilasters: □ No □ Yes Size _____
Glass Block: □ No □ Yes Size _____ Type _____
Special Block _____
Describe _____
Glazed Concrete Block: □ No □ Yes □ Solid □ Hollow Type _____
□ Non Reinforced □ Reinforced _____ Strips Every _____ Course.
Describe _____
Gypsum Block: □ No □ Yes □ Solid □ Hollow Thickness _____ Describe _____

Grouting: □ No □ Yes □ Block Cores □ Bond Beams □ Cavity Walls □ Door Frames □ Lintels □ _____
Describe _____
Insulation: □ No □ Yes □ Board □ Poured □ Sprayed Material _____
Thickness _____ _____ Describe _____
Solar Screen: □ No □ Yes Describe _____
Special Block: □ No □ Yes Describe _____
□ Parge Block □ Clean Cavity □ Spandrel Flashing _____
Special Considerations: _____
Ceramic Veneer: □ No □ Yes Describe _____
Structural Facing Tile: □ No □ Yes □ 6T Series □ 8W Series □ Other _____
Describe _____
Terra Cotta: □ No □ Yes □ Floors □ Partitions □ Fireproofing □ Load Bearing □ Non Load Bearing
Describe _____
Ashlar Stone: □ No □ Yes Type _____ Thickness _____
Describe _____
Rubble Stone: □ No □ Yes □ Coarsed □ Uncoarsed Type _____ Thickness _____
Describe _____
Cut Stone: □ No □ Yes □ Granite □ Limestone □ Marble □ Sand Stone □ Slate □ _____
□ Base _____ □ Columns _____ □ Coping _____
□ Curbs _____ □ Facing Panels _____
□ Flooring _____ □ Showers _____
□ Soffits _____ □ Stair Treads _____
□ Stairs _____ □ Thresholds _____
□ Window Sills _____ □ Window Stools _____
□ _____
Simulated Stone: □ No □ Yes Material _____ Describe _____
Special Stone _____
General Notes: Masonry _____
_____
_____

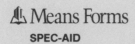**Means Forms**

SPEC-AID

DATE _____

DIVISION 5: METALS

PROJECT _____   LOCATION _____

Design Criteria: In Division 1 _____
   Typical Bay Spacings: _____
   _____
   Floor to Ceiling Heights: _____
   Beam Depths: _____
   Roof Slope: ☐ Flat   ☐ Other _____
   Eave Height: _____

Anchor Bolts: ☐ No   ☐ By Others   ☐ Yes   Describe _____ Number _____

Base Plates: ☐ No   ☐ By Others   ☐ Yes   Describe _____
   _____ Number _____

Metal Decking Floors: ☐ No   ☐ By Others   ☐ Cellular   ☐ Non Cellular   ☐ Painted   ☐ Galv.   Depth _____
   ☐ Acoustical   ☐ Ventilating   Gauge _____ Describe _____

   Roof Deck: ☐ No   ☐ By Others   ☐ Cellular   ☐ Non Cellular   ☐ Painted   ☐ Galv.   Depth _____
   ☐ Acoustical   ☐ Ventilating   Gauge _____ Describe _____

Structural System: ☐ Wall Bearing   ☐ Free Standing   ☐ Simple Spans   ☐ Continuous Spans _____
   ☐ Conventional Design   ☐ Plastic Design   ☐ Field Welded   ☐ Field Bolted   ☐ Composite Design
   ☐ Other _____
   Type Steel _____ Grade _____
   Estimated Weights: Beams _____ Roof Frames _____ Adjustable Spandrel Angles _____
   Girders _____ Girts _____ Hanger Pods _____ Bracing _____
   Columns _____ Connections _____ Other _____
   Paint: Shop   ☐ No   ☐ Yes _____ Coats   Material _____
   Field Paint   ☐ No   ☐ Yes   ☐ Brush   ☐ Roller   ☐ Spray _____ Coats   Material _____
   Galvanizing: ☐ No   ☐ Yes   Thickness _____
   Other: _____

Fireproofing: ☐ No   ☐ Yes   ☐ Beams   ☐ Columns   ☐ Decks   ☐ Other _____ Rating _____ Hr.
   ☐ Concrete Encasement   ☐ Spray   ☐ Plaster   ☐ Drywall   ☐ _____
   Describe _____

Open Web Joists: ☐ No   ☐ Yes   ☐ H Series   ☐ LH Series   ☐ _____
   Estimated Weights: _____
   Bridging: ☐ No   ☐ Yes   ☐ Bolted   ☐ Welded   ☐ Pod Bridging _____
   Paint: Shop   ☐ Standard   ☐ Special _____ Coats   Field Paint   ☐ No   ☐ Yes   Describe _____

Light Gauge Joists: ☐ No   ☐ Yes   Describe _____

Light Gauge Framing: ☐ No   ☐ Yes   Describe _____

Special Considerations: _____
_____
_____
_____

Fasteners: Expansion Bolts _____ High Strength Bolts _____
   Machine Screws _____ Machinery Anchors _____
   Nails _____ Roof Bolts _____
   Sheet Metal Screws _____ Studs _____
   Timber Connectors _____ Toggle Bolts _____
   Welded Studs _____ Other _____
_____
_____

SPEC-AID

DIVISION 5: METALS

Area Walls: ☐ No ☐ Yes _____ Gratings _____ Caps _____

Bumper Rails: ☐ No ☐ Yes _____

Canopy Framing: ☐ No ☐ Yes _____

Checkered Plate: ☐ No ☐ Trench Covers ☐ Pit Covers ☐ Platforms ☐ _____

Columns: ☐ No ☐ Aluminum ☐ Steel ☐ Square ☐ Rectangular ☐ Round ☐ _____

Construction Castings: ☐ No ☐ Chimney Specialties _____ ☐ Column Bases _____

☐ Manhole Covers _____ ☐ Wheel Guards _____

☐ _____

Corner Guards: ☐ No ☐ Yes _____

Crane Rail: ☐ No ☐ Yes _____

Curb Angles: ☐ No ☐ Straight ☐ Curved _____

Decorative Covering: ☐ No ☐ Stock Sections ☐ Custom Sections _____

Doors _____ Walls _____

Door Frames: ☐ No ☐ Yes _____ ☐ Protection: ☐ No ☐ Yes _____

Expansion Joints Ceilings: ☐ No ☐ Yes _____ Cover Plates: ☐ No ☐ Yes _____

Floors: ☐ No ☐ Yes _____ Cover Plates: ☐ No ☐ Yes _____

Walls: ☐ No ☐ Yes _____ Cover Plates: ☐ No ☐ Yes _____

Fire Escape: ☐ No ☐ Yes _____ Size _____

Stairs _____ Ladders _____ Cantilever _____

Floor Grating: ☐ No ☐ Aluminum ☐ Steel ☐ Fiberglass ☐ Platforms ☐ Stairs ☐ _____

Type _____ Weight _____

Special Finish _____

Ladders: ☐ No ☐ Aluminum ☐ Steel ☐ _____

☐ With Cage ☐ No Cage _____ ☐ Inclined Type _____

Lamp Posts: ☐ No ☐ Yes _____

Lintels: ☐ No ☐ Yes ☐ Plain ☐ Built-up ☐ Painted ☐ Galvanized _____

_____

Louvers: ☐ No ☐ Yes _____

Manhole Covers: ☐ No ☐ Yes _____

Mat Frames: ☐ No ☐ Yes _____

Overhead Supports: ☐ No ☐ Toilet ☐ Partitions ☐ _____

Pipe Bumpers: ☐ No ☐ Yes _____

Pipe Supports: ☐ No ☐ Yes _____

Railings: ☐ No ☐ Yes ☐ Aluminum ☐ Steel ☐ Pipe ☐ _____

Balconies: ☐ No ☐ Yes _____

Stairs: ☐ No ☐ Yes _____

Wall: ☐ No ☐ Yes _____

Solar Screens: ☐ No ☐ Yes _____

Stairs: ☐ No ☐ Yes ☐ Aluminum ☐ Steel ☐ Stock ☐ Custom ☐ _____

Size _____ Landings _____

Spiral: ☐ No ☐ Aluminum ☐ Steel ☐ Stock ☐ Custom ☐ _____

Pre-erected: ☐ No ☐ Yes _____

Stair Treads: ☐ No ☐ Yes _____

Trench Covers: ☐ No ☐ Yes _____

Weather Vanes: ☐ No ☐ Yes _____

Window Guards: ☐ No ☐ Bars ☐ Woven Wire ☐ _____

Wire: ☐ No ☐ Yes _____

Wire Rope: ☐ No ☐ Yes _____

Special Considerations: _____

_____

_____

_____

SPEC-AID
DIVISION 6: CARPENTRY

DATE _____

PROJECT _____ LOCATION _____

Framing: Type Wood _____ Fiber Stress _____ psi.

Beams: ☐ Single ☐ Built Up _____ Grade _____

Bracing: ☐ No ☐ Let In ☐ _____

Bridging: ☐ Steel ☐ Wood _____

Canopy Framing: _____

Columns _____ Fiber Stress _____ psi.

Door Bucks: ☐ No ☐ Treated ☐ Untreated _____

Floor Planks _____ Grade _____

Furring: ☐ Metal ☐ Wood _____

Grounds: ☐ No ☐ Casework ☐ Plaster ☐ On Wood ☐ On Masonry _____

Joists: ☐ No ☐ Floor ☐ Ceiling _____ Grade _____

Ledgers: ☐ No ☐ Bolted ☐ Nailed _____

Lumber Treatment: ☐ No ☐ Creosote ☐ Salt Treated ☐ Fire Retardant _____
☐ Kiln Dry _____

Nailers: ☐ No ☐ Treated ☐ Untreated _____

Plates: _____ **Platform Framing:** _____

Plywood Treatment: ☐ No ☐ Salt Treated ☐ Fire Retardant _____

Posts & Girts: _____

Rafters: ☐ No ☐ Ordinary ☐ Hip _____

Roof Cants: ☐ No ☐ Yes _____ **Roof Curbs:** ☐ No ☐ Yes _____

Roof Decks: ☐ No ☐ Yes _____ Inches Thick

Roof Purlins: ☐ No ☐ Yes _____

Roof Trusses: ☐ No ☐ Timber Connectors ☐ Nailed ☐ Glued Spaced _____ O.C. Span _____ feet

Sheathing, Roof: ☐ No ☐ Plywood ☐ Boards ☐ Wood Fiber ☐ Gypsum _____

Wall: ☐ No ☐ Plywood ☐ Boards ☐ Wood Fiber ☐ Gypsum _____

Siding Hardboard: ☐ Plain ☐ Primed ☐ Stained _____

Particle Board: _____ Wood Fiber _____

Plywood: ☐ Cedar ☐ Fir ☐ Redwood ☐ Marine ☐ Natural ☐ Stained ☐ Plastic Faced ☐ _____

Wood: ☐ Cedar ☐ Redwood ☐ White Pine ☐ Bevel ☐ Board & Batten ☐ Channel ☐ T & C ☐ Shiplap _____
☐ Natural ☐ Stained _____

Sills: _____ **Sleepers:** _____

Soffits: ☐ No ☐ Open ☐ Vented ☐ Plywood _____

Stressed Skin Plywood Box Beams: _____ Depth _____

Floor Panels: ☐ No ☐ Yes _____ Depth _____

Roof Panels: ☐ No ☐ Straight ☐ Curved _____ Depth _____

Folded Plate: ☐ No ☐ Yes _____ Depth _____

Studs: ☐ No ☐ Yes _____ Grade _____

Subfloor: ☐ No ☐ Plywood ☐ Boards ☐ Wood Fiber ☐ _____

Suspended Ceiling Framing: ☐ No ☐ Yes _____

Underlayment: ☐ No ☐ Particle Board ☐ Plywood ☐ Wood Fiber ☐ Hardboard _____

Special Considerations: _____

Laminated Framing: ☐ Beams ☐ Straight ☐ Curved _____ Span _____
☐ Bowstring Trusses ☐ Radial Arch ☐ Tudor Arch ☐ Columns _____
Span _____ Height _____
☐ Industrial Grade ☐ Premium Grade ☐ Exterior Glue ☐ Stain ☐ Varnish ☐ Treated ☐ _____

Laminated Roof Deck: ☐ No ☐ Yes _____ Thickness _____

Special Considerations: _____

DIVISION 6: CARPENTRY

Base: ☐ No ☐ One Piece ☐ Built up ☐ Pine ☐ Hardwood _____

Cabinets: ☐ No ☐ Corner ☐ Kitchen ☐ Toilet Room ☐ Other _____

    ☐ Stock ☐ Custom _____

    ☐ Unfinished ☐ Prefinished _____

    Base Cabinets: ☐ Softwood ☐ Hardwood ☐ Drawer Units _____

    Wall Cabinets: ☐ Softwood ☐ Hardwood _____

    Tall Cabinets: ☐ Softwood ☐ Hardwood _____

    Special: _____

Casings: ☐ No ☐ Doors ☐ Windows ☐ Beams ☐ Others _____

    ☐ Softwood ☐ Hardwood _____

Ceiling Beams: ☐ No ☐ Cedar ☐ Pine ☐ Fir ☐ Plastic _____

Chair Rail: ☐ No ☐ Pine ☐ Other _____

Closets: ☐ No ☐ Pole ☐ Shelf ☐ Prefabricated _____

Columns: ☐ No ☐ Square ☐ Round ☐ Solid ☐ Built up ☐ Hollow ☐ Tapered _____

    Diameter _____ Height _____

Convector Covers: ☐ No ☐ Yes _____

Cornice: ☐ No ☐ 1 Piece ☐ 2 Piece ☐ 3 Piece ☐ Pine ☐ Cedar ☐ Other _____

Counter Tops: ☐ No ☐ Plastic ☐ Ceramic Tile ☐ Marble ☐ Suede Finish ☐ Other _____

    ☐ Stock ☐ Custom _____

    ☐ No Splash ☐ Square Splash ☐ Cover Splash _____

    ☐ Self Edge ☐ Stainless Edge ☐ Aluminum Edge _____

    Special _____

Cupolas: ☐ No ☐ Stock ☐ Custom ☐ Wood ☐ Fiberglass ☐ Square ☐ Octagonal Size _____

    ☐ Aluminum Roof ☐ Copper Roof ☐ Other _____

Doors and Frames: See Division 8 _____

Door Moldings: ☐ No ☐ Yes _____

Door Trim: ☐ No ☐ Yes _____

Fireplace Mantels: ☐ No ☐ Beams ☐ Moldings _____

    Size _____

Moldings: ☐ No ☐ Softwood ☐ Hardwood ☐ Metal ☐ Other _____

Paneling Hardboard: ☐ No ☐ Tempered ☐ Untempered ☐ Pegboard ☐ Plastic Faced _____

    Plywood, Unfinished: ☐ No ☐ Veneer Core ☐ Lumber Core Grade _____ Thick _____

    Plywood, Prefinished: ☐ No ☐ Stock ☐ Architectural Finish _____

    Size _____

    Wood Boards: ☐ No ☐ Softwood ☐ Hardwood _____

Railings: ☐ No ☐ Stock ☐ Custom ☐ Softwood ☐ Hardwood _____

    ☐ Stairs ☐ Balcony ☐ Porch ☐ Wall ☐ Other _____

Shelving: ☐ No ☐ Prefinished ☐ Unfinished ☐ Stock ☐ Custom ☐ Plywood ☐ Particle Board ☐ Boards _____

    ☐ Book Shelves _____ ☐ Linen Shelves _____

    ☐ Storage Shelves _____ ☐ Other _____

Stairs: ☐ No ☐ Prefabricated ☐ Built in Place ☐ Softwood ☐ Hardwood _____

    ☐ Box ☐ Open ☐ Circular _____

Thresholds: ☐ No ☐ Interior ☐ Exterior _____

Wainscot: ☐ No ☐ Boards ☐ Plywood ☐ Moldings _____

Windows and Frames: See Division 8 _____

Window Trim: ☐ No ☐ Yes _____

Special Considerations: _____

DIVISION 7: MOISTURE PROTECTION

PROJECT _____   LOCATION _____

Bentonite: ☐ No   ☐ Panels   ☐ Granular _____

Bituminous Coating: ☐ No   ☐ Brushed   ☐ Sprayed   ☐ Troweled   ☐ 1 Coat   ☐ 2 Coat   ☐ Protective Board _____

Building Paper: ☐ No   ☐ Asphalt   ☐ Polyethylene   ☐ Rosin   ☐ Kraft   ☐ Foil Backed   ☐ _____

☐ Roof Deck Vapor Barrier _____

Caulking: ☐ No   ☐ Gun Grade   ☐ Knife Grade   ☐ Plain   ☐ Colors _____

☐ Doors   ☐ Windows   ☐ _____

Cementitious: ☐ No   ☐ 1 Coat   ☐ 2 Coat   Thickness _____ Inches   Mix _____

Control Joints: _____   **Expansion Joints:** _____

Elastomeric Waterproofing: ☐ No   ☐ EPDM   ☐ Neoprene   ☐ PVC   ☐ Urethane   ☐ _____

Liquid Waterproofing: ☐ No   ☐ Silicone   ☐ Stearate   ☐ _____

Membrane Waterproofing: ☐ 1 Ply   ☐ 2 Ply   ☐ 3 Ply   ☐ Felt   ☐ Fabric   ☐ Elastomeric   ☐ _____

Metallic Coating: ☐ No   ☐ Walls _____ in. Thick   ☐ Floors _____ in. Thick _____

Preformed Vapor Barrier: ☐ No   ☐ Yes _____

Sealants: ☐ No   ☐ Butyl   ☐ Polysulfide   ☐ PVC   ☐ Urethane   ☐ _____

☐ Doors   ☐ Windows   ☐ _____

Special Waterproofing _____

_____

Building Insulation: Rigid: ☐ No   ☐ Fiberglass   ☐ Polystyrene   ☐ Urethane   ☐ _____

Non Rigid: ☐ No   ☐ Fiberglass   ☐ Mineral Fiber   ☐ Vermiculite   ☐ Perlite   ☐ _____

Form Board: ☐ No   ☐ Acoustical   ☐ Asbestos Cement   ☐ Fiberglass   ☐ Gypsum   ☐ Mineral Fiber   ☐ Wood Fiber

☐ Other _____   ☐ Sub Purlins _____ Span _____

Masonry Insulation: ☐ No   ☐ Cavity Wall   ☐ Block Cores   ☐ Poured   ☐ Foamed Type _____

Perimeter Insulation: ☐ No   ☐ Yes   Type _____   Thickness _____

Roof Deck Insulation: ☐ No   ☐ Fiberboard   ☐ Fiberglass   ☐ Foamglass   ☐ Polystyrene   ☐ Urethane   ☐ _____

Thickness _____   ☐ **Cants** _____ Size _____

Sprayed: ☐ No   ☐ Fibrous   ☐ Cementitious   ☐ Urethane   ☐ _____

Special Insulation _____

_____

Shingles: Aluminum: ☐ No   ☐ Yes _____   Asbestos: ☐ No   ☐ Yes _____

Asphalt: ☐ No   ☐ Class C   ☐ Class A   ☐ _____   Weight _____ lb. per Sq. _____

Clay Tile: ☐ No   ☐ Plain   ☐ Glazed   ☐ Spanish   ☐ _____   Weight _____ lb. per Sq. _____

Concrete Tile: ☐ No   ☐ Yes _____   Porcelain Enamel: ☐ No   ☐ Yes _____

Slate: ☐ No   ☐ Yes   Type _____   Color _____ Exposure _____

Wood: ☐ No   ☐ Roofing   ☐ Siding   ☐ Fire Retardant   Type ____ Grade _____ Exposure _____

Shingle Underlayment: ☐ No   ☐ Asbestos   ☐ Asphalt   ☐ _____   Weight _____

Special Shingles: _____

_____

Aluminum: ☐ No   ☐ Roofing   ☐ Siding   ☐ Painted   ☐ Insulated   ☐ Sandwich   ☐ _____

Thickness _____ Type _____

Asbestos Cement: ☐ No   ☐ Roofing   ☐ Siding   ☐ Flat   ☐ Corrugated   ☐ Natural   ☐ Painted   ☐ Sandwich

☐ Fire Rated   Thickness _____ Type _____

Epoxy Panels: ☐ No   ☐ Solid   ☐ Plywood Back   ☐ Hardboard Back   ☐ Exposed Aggregate   ☐ _____

Fiberglass Panels: ☐ No   ☐ Roofing   ☐ Siding   ☐ Flat   ☐ Corrugated   ☐ _____ Thickness _____

Metal Facing Panels: ☐ No   ☐ Field Assembled   ☐ Factory Made   Insulation _____

Outside Face _____   Inside Face _____

Protected Metal: ☐ No   ☐ Roofing   ☐ Siding   Type _____   Gauge _____

Steel: ☐ No   ☐ Roofing   ☐ Siding   ☐ Painted   ☐ Galvanized   ☐ Insulated   ☐ Sandwich   ☐ _____

Type _____   Gauge _____

Vinyl Siding: ☐ No   ☐ Plain   ☐ Insulated   Type _____

Special Roofing & Siding: _____

_____

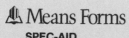
Means Forms
SPEC-AID

DIVISION 7: MOISTURE PROTECTION

Built Up Roofing: ☐ No ☐ Tar & Gravel ☐ Asphalt & Gravel ☐ Felt ☐ Mineral Surface ☐ Aggregate
☐ 1 Ply ☐ 2 Ply ☐ 3 Ply ☐ 4 Ply ☐ 5 Ply ☐ Bonded _____ years Roof Pitch _____ Type Deck _____
Underlayment: ☐ No ☐ Rosin Paper ☐ Vapor Barrier ☐ _____

Elastic Sheet Roofing: ☐ No ☐ Butyl ☐ Neoprene ☐ _____ Thickness _____
Describe _____

Fluid Applied Roofing: ☐ No ☐ Hypalon Neoprene ☐ Silicone ☐ Vinyl ☐ _____ Thickness _____
Describe _____

Roll Roofing: ☐ No ☐ Smooth ☐ Granular _____ Weight _____ lbs. per Sq.
Special Membrane Roofing: _____

Downspouts: ☐ No ☐ Aluminum ☐ Copper ☐ Lead Coated Copper ☐ Galvanized Steel ☐ Stainless Steel
☐ Steel Pipe ☐ Vinyl ☐ Zinc Alloy ☐ Stock ☐ Custom ☐ _____ Size _____
Describe _____

Expansion Joints: ☐ No ☐ Roof ☐ Walls ☐ No Curbs ☐ Curbs ☐ Rubber ☐ Metallic ☐ _____
Describe _____

Fascia: ☐ No ☐ Yes Describe _____ Thickness _____

Flashing: ☐ No ☐ Aluminum ☐ Asphalt ☐ Copper ☐ Fabric ☐ Lead ☐ Lead Coated Copper ☐ PVC ☐ Rubber
☐ Stainless Steel ☐ Terne ☐ Zinc Alloy ☐ Paper Backed ☐ Mastic Backed ☐ Fabric Backed ☐ _____
Describe _____ Thickness _____

Gravel Stop: ☐ No ☐ Aluminum ☐ Copper ☐ PVC ☐ Stainless Steel ☐ _____
☐ With Fascia ☐ No Fascia ☐ Natural ☐ Painted Thickness _____ Face Height _____

Gutters: ☐ No ☐ Aluminum ☐ Copper ☐ Lead Coated Copper ☐ Galvanized Steel ☐ Stainless Steel _____
☐ Vinyl ☐ Wood ☐ Zinc Alloy ☐ _____ Thickness _____
☐ Box Type ☐ K Type ☐ Half Round ☐ Stock ☐ Custom ☐ _____ Size _____

Louvers: ☐ No ☐ Yes _____

Mansard: ☐ No ☐ Yes _____ Thickness _____

Metal Roofing: ☐ No ☐ Copper ☐ Copper Bearing Steel ☐ Lead ☐ Lead Coated Copper ☐ Stainless Steel
☐ Terne ☐ Zinc Alloy ☐ _____ Size _____ Thickness _____
☐ Standing Seam ☐ Flat Seam ☐ Batten Seam ☐ _____ Weight _____ lbs. per Sq.

Underlayment: ☐ No ☐ 15 lb. Felt ☐ 30 lb. Felt ☐ Rosin Paper ☐ _____

Reglet: ☐ No ☐ Aluminum ☐ Copper ☐ Galvanized Steel ☐ Stainless Steel ☐ Zinc Alloy ☐ _____
Thickness _____ Counter Flashing: ☐ No ☐ Yes _____ Thickness _____

Soffit: ☐ No ☐ Yes _____ Thickness _____
Special Sheet Metal Work: _____

Ceiling Hatches: ☐ No ☐ Steel ☐ Galvanized ☐ Painted ☐ Aluminum ☐ _____
Size _____

Roof Drains: ☐ No ☐ In Plumbing ☐ Yes _____

Roof Hatches: ☐ No ☐ Steel ☐ Galvanized ☐ Painted ☐ Aluminum ☐ _____
☐ Insulated ☐ Not Insulated ☐ With Curbs ☐ No Curbs ☐ _____ Size _____

Smoke Hatches: ☐ No ☐ Yes _____

Snow Guards: ☐ No ☐ Yes _____

Skylights: ☐ No ☐ Domes ☐ Vaulted ☐ Ridge Units ☐ Field Fabricated ☐ Glass ☐ Plastic ☐ Single
☐ Double ☐ Sandwich Panels ☐ With Curbs ☐ No Curbs ☐ _____ Size _____

Smoke Vents: ☐ No ☐ Yes _____

Skyroofs: ☐ No ☐ Yes _____

Ventilators: ☐ No ☐ In Ventilating ☐ Stationary ☐ Spinners ☐ Motorized _____
Special Roof Accessories _____

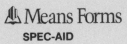

SPEC-AID

DATE _____

DIVISION 8: DOORS, WINDOWS & GLASS

PROJECT _____ LOCATION _____

Hollow Metal Frames: ☐ No ☐ Baked Enamel ☐ Galvanized ☐ Porcelain Enamel _____

Hollow Metal Doors: ☐ No _____ ☐ Core _____ ☐ Labeled _____

Aluminum Frames: ☐ No ☐ Clear ☐ Bronze ☐ Black _____

Aluminum Doors and Frames: ☐ No ☐ Yes _____ Frames _____

Wood Frames: ☐ No ☐ Exterior ☐ Interior ☐ Custom ☐ With Sill ☐ Vinyl Covered ☐ Pine ☐ Oak

Wood Doors: ☐ No _____ Core ☐ Labeled _____ Frames _____

Interior Door Frames: ☐ No ☐ Aluminum ☐ Hollow Metal ☐ Steel ☐ Wood ☐ Prehung ☐ Stock

☐ Custom ☐ _____

Custom Doors: ☐ No ☐ Swing ☐ Bi-Passing ☐ Bi-Folding _____ Frames _____

Accordion Folding Doors: ☐ No ☐ Yes _____ Frames _____

Acoustical Doors: ☐ No ☐ Yes _____ Decibels _____ Frames _____

Cold Storage: ☐ No ☐ Manual ☐ Power ☐ Sliding ☐ Hinged _____

Counter Doors: ☐ No ☐ Aluminum ☐ Steel ☐ Wood _____ Frames _____

Dark Room Doors: ☐ N0 ☐ Revolving ☐ 2 Way ☐ 3 Way _____

Floor Opening Doors: ☐ No ☐ Aluminum ☐ Steel ☐ Single ☐ Double ☐ Commercial ☐ Industrial ____

Glass Doors: ☐ No ☐ Sliding ☐ Swing _____ Frames _____

Hangar Doors: ☐ No ☐ Bi-Fold ☐ Other ☐ Electric _____

Jalousie Doors: ☐ No ☐ Plain Glass ☐ Tempered Glass _____

Kalamein: ☐ No ☐ Yes ☐ Labeled _____ Frames _____

Kennel Doors: ☐ No ☐ 2 Way Swing _____

Overhead Doors: ☐ No ☐ Regular Duty ☐ Heavy Duty ☐ Stock ☐ Custom ☐ One Piece ☐ Sectional

☐ Manual ☐ Electric ☐ Aluminum ☐ Fiberglass ☐ Steel ☐ Wood ☐ Hardboard ☐ Commercial

☐ Residential Size _____

Rolling Doors Exterior: ☐ No ☐ Manual ☐ Electric ☐ Labeled _____

Rolling Doors Interior: ☐ No ☐ Manual ☐ Electric ☐ Labeled _____ Frames _____

Rolling Grilles: ☐ No ☐ Manual ☐ Electric ☐ Aluminum ☐ Steel _____

Service Door Frames: ☐ No ☐ Aluminum ☐ Hollow Metal ☐ Steel ☐ Wood ☐ Stock ☐ Custom ____

Service Doors: ☐ No ☐ Stock ☐ Custom _____ ☐ Transoms _____ ☐ Sidelights _____

☐ Aluminum _____ ☐ Hollow Metal _____ ☐ Core _____ ☐ Kalamein _____

☐ Steel _____ ☐ Wood _____ ☐ Core _____ ☐ Labeled _____ ☐ Special Finish ____

Shock Absorbing Doors: ☐ No ☐ Flexible ☐ Rigid _____ Frames _____

Sliding Doors: ☐ No ☐ Glazed ☐ Unglazed Aluminum ☐ Steel ☐ Wood _____

Swing Doors: ☐ No ☐ Single ☐ Double _____

Telescoping Door: ☐ No ☐ Manual ☐ Electric _____

Tinclad Doors: ☐ No ☐ Manual ☐ Electric _____

Vault Front Doors: ☐ No ☐ Stainless Steel ☐ Time Lock ☐ 1 Hr. Test ☐ 2 Hr. Test ☐ 4 Hr. Test ____

Special Exterior Doors: ☐ No ☐ Yes _____

Special Interior Doors: ☐ No ☐ Yes _____

Balanced Doors: ☐ No ☐ Economy ☐ Premium ☐ Aluminum ☐ Stainless Steel _____

Revolving Doors: ☐ No ☐ Stock ☐ Custom ☐ Manual ☐ Electric ☐ Diameter _____

Entrance Units: ☐ No ☐ Aluminum ☐ Bronze ☐ Glass ☐ Hollow Metal ☐ Stainless Steel ☐ Wood ☐ Steel

☐ Stock ☐ Custom ☐ Balanced ☐ Sidelights ☐ Transoms Special Finish _____

Entrance Frames: ☐ No ☐ Aluminum ☐ Hollow Metal ☐ Steel ☐ Wood ☐ Stainless Steel ☐ Stock

☐ Custom _____

Store Fronts: ☐ No ☐ Sliding ☐ Fixed ☐ Institutional Grade ☐ Monumental Grade ☐ Commercial Grade

Windows: _____ % of Exterior Walls _____

Projected: ☐ No ☐ Glazed ☐ Unglazed ☐ Aluminum ☐ Steel ☐ Wood _____

Single Hung: ☐ No ☐ Glazed ☐ Unglazed ☐ Aluminum ☐ Steel ☐ Wood _____

Sliding: ☐ No ☐ Glazed ☐ Unglazed ☐ Aluminum ☐ Steel ☐ Wood _____

Security Windows: ☐ No ☐ Yes _____

⚓ Means Forms

SPEC-AID

DIVISION 8: DOORS, WINDOWS & GLASS

Casement: ☐ No ☐ Fixed _____% Vented ☐ Aluminum ☐ Steel ☐ Wood _____

Picture Window: ☐ No ☐ Glazed ☐ Unglazed ☐ Aluminum ☐ Steel ☐ Wood _____

Double Hung: ☐ No ☐ Glazed ☐ Unglazed ☐ Aluminum ☐ Steel ☐ Wood _____

Special Windows: ☐ No ☐ Yes _____

Screens: ☐ No ☐ Aluminum ☐ Steel ☐ Wood _____

Finish Hardware Allowance: ☐ No ☐ Yes _____

Exterior Doors _____

Interior Doors _____

Automatic Openers: ☐ No ☐ 1 Way ☐ 2 Way ☐ Double Door ☐ Activating Carpet

Automatic Operators: ☐ No ☐ Sliding ☐ Swing ☐ Controls _____

Bumper Plates: ☐ No ☐ U Channel ☐ Teardrop _____

Door Closers: ☐ No ☐ Regular ☐ Fusible Link ☐ Concealed ☐ Heavy Use _____

Door Stops: ☐ No ☐ Yes

Floor Checks: ☐ No ☐ Single Acting ☐ Double Acting _____

Hinges: ☐ No ☐ Butt ☐ Pivot ☐ Spring ☐ Frequency _____

Kick Plates: ☐ No ☐ Yes _____

Lock Set: ☐ No ☐ Cylindrical ☐ Mortise ☐ Heavy Duty ☐ Commercial ☐ Residential _____

Panic Device: ☐ No ☐ Yes ☐ Exit Only ☐ Exit & Entrance _____

Push-Pull Device: ☐ No ☐ Yes ☐ Bronze ☐ Aluminum ☐ Other _____

Cabinet Hardware: ☐ No ☐ Yes _____

Window Hardware: ☐ No ☐ Yes _____

Special Hardware: ☐ No ☐ Yes _____

Threshold: ☐ No ☐ Yes _____

Weather Stripping Doors: ☐ No ☐ Zinc ☐ Bronze ☐ Stainless Steel ☐ Spring Type ☐ Extruded Sections

Windows: ☐ No ☐ Zinc ☐ Bronze _____

Acoustical Glass: ☐ No ☐ Yes _____ Thickness _____

Faceted Glass: ☐ No ☐ Yes _____ Thickness _____

Glazing: ☐ No ☐ Putty ☐ Flush ☐ Bead ☐ Gasket ☐ Butt ☐ Riglet ☐ _____

Insulated Glass: ☐ No ☐ Standard ☐ Non-Standard _____ Thickness _____

Laminated Glass: ☐ No ☐ Yes _____ Thickness _____

Mirrors: ☐ No ☐ Plate ☐ Sheet ☐ Transparent ☐ Incl. Frames ☐ No Frames ☐ _____

Door Type _____ Wall Type _____

Obscure Glass: ☐ No ☐ Yes _____ Thickness _____

Plate Glass: ☐ No ☐ Clear ☐ Tinted ☐ Tempered _____ Thickness _____

Plexiglass: ☐ No ☐ Masked ☐ Unmasked _____ Thickness _____

Polycarbonate: ☐ No ☐ Masked ☐ Unmasked _____ Thickness _____

Reflective: ☐ No ☐ Clear ☐ Tinted _____ Thickness _____

Sand Blasted: ☐ No ☐ Yes _____ Thickness _____

Sheet or Float Glass: ☐ No ☐ Clear ☐ Gray _____ Thickness _____

Spandrel Glass: ☐ No ☐ Plain ☐ Insulated ☐ Sandwich _____ Thickness _____

Stained Glass: ☐ No ☐ Yes _____

Vinyl Glazing: ☐ No ☐ Yes _____ Thickness _____

Window Glass: ☐ No ☐ DSA ☐ DBS ☐ Tempered _____ Thickness _____

Wire Glass: ☐ No ☐ Yes _____ Thickness _____

Special Glazing: ☐ No ☐ Yes _____

Curtain Walls: ☐ No ☐ Yes _____

Window Walls: ☐ No ☐ Yes _____

PROJECT _____   LOCATION _____

Furring: Ceiling: ☐ No   ☐ Wired Direct   ☐ Suspended _____

Partitions: ☐ No   ☐ Load Bearing   ☐ Non Load Bearing _____ Thickness _____

Walls: ☐ No   ☐ Yes _____

Gypsum Lath: ☐ No   ☐ Walls   ☐ Ceilings   ☐ Regular   ☐ Foil Faced   ☐ Fire Resistant   ☐ Moisture Resistant _____

_____ Thickness _____

Metal Lath: ☐ No   ☐ Diamond   ☐ Rib   ☐ _____ Weight _____

☐ Painted   ☐ Galvanized   ☐ Paper Backed _____

☐ Walls   ☐ Ceilings   ☐ Suspended   ☐ Partitions   ☐ Load Bearing   ☐ Non Load Bearing _____

_____

Drywall Finishes: ☐ Taped & Finished   ☐ Thin Coat Plaster   ☐ Prime Coat   ☐ Electric Heat Compound   ☐ _____

Mountings: ☐ Nailed   ☐ Screwed   ☐ Laminated   ☐ Clips   ☐ _____

Beams: ☐ No _____ Layers _____ Thickness _____

Ceilings: ☐ No   ☐ Standard   ☐ Fire Resistant   ☐ Water Resistance _____ Thickness _____

Columns: ☐ No _____ Layers _____ Thickness _____

Partitions: ☐ No   ☐ Wood Studs   ☐ Steel Studs _____ Layers _____ Thickness _____

Prefinished: ☐ No   ☐ Standard   ☐ Fire Resistance _____ Thickness _____

Soffits: ☐ No _____ Layers _____ Thickness _____

Sound Deading Board: ☐ No   Type _____ Thickness _____

Walls: ☐ No _____ Layers _____ Thickness _____

Plaster Finishes: ☐ 1 Coat   ☐ 2 Coat   ☐ 3 Coat   ☐ Gypsum   ☐ Perlite   ☐ Vermiculite   ☐ Wood   ☐ _____

Beams: ☐ No _____ Ceilings: ☐ _____

Columns: ☐ No _____ Soffits: ☐ No _____

Partitions: ☐ No   ☐ Wood Studs   ☐ Steel Studs   ☐ Solid   ☐ Hollow _____

Walls: ☐ No _____

Special Plaster: _____

Sprayed Acoustical: ☐ No   ☐ Yes _____ Thickness _____

Fireproofing: ☐ No   ☐ Yes _____ Thickness _____

Stucco: ☐ No   ☐ On Mesh   ☐ Masonry _____

Cast Stone: ☐ No   ☐ Glazed   ☐ Unglazed   ☐ Waxed _____ Thickness _____

Ceramic Tile Base: ☐ No   ☐ Cove   ☐ Sanitary   ☐ _____ Set _____ Height _____

Floors: ☐ No   ☐ _____ Set   ☐ Natural Clay   ☐ Porcelain   ☐ Conductive _____ Color Group _____

Walls: ☐ No   ☐ _____ Set   ☐ Interior   ☐ Exterior   ☐ Glazed   ☐ Crystalline Glazed   ☐ _____

☐ Unmounted   ☐ Backmounted _____

Panels: ☐ No   ☐ Yes _____

Glass Mosaics: ☐ No   ☐ Yes _____ Color Group _____

Metal Tile: ☐ No   ☐ Aluminum   ☐ Copper   ☐ Stainless Steel _____

Plastic Tile: ☐ No   ☐ Yes _____ Thickness _____

Quarry Tile Base: ☐ No   ☐ Cove   ☐ Sanitary _____ Height _____

Floor   ☐ No _____ Set   Size _____ Color _____

Stairs: ☐ No   ☐ Treads   ☐ Risers _____

Wainscot: ☐ No _____ Set   Size _____

Cast In Place Terrazzo Base: ☐ No   ☐ Yes _____ Curb: ☐ No   ☐ Yes _____

Floor: ☐ No   ☐ Bonded   ☐ Unbonded   ☐ Gray Cement   ☐ White Cement   ☐ Conventional   ☐ Venetian _____

☐ Conductive   ☐ Monolithic   ☐ Epoxy   ☐ _____

Divider Strips: ☐ No   ☐ Brass   ☐ Zinc _____ Spacing _____

Stairs: ☐ No   ☐ Yes _____ Wainscot: ☐ No   ☐ Yes _____

Precast Terrazzo Base: ☐ No   ☐ Yes   Curb: ☐ No   ☐ Yes _____

Floor Tiles: ☐ No   Size _____ Thickness _____

Stairs: ☐ No   ☐ Treads   ☐ Risers   ☐ Stringers   ☐ Landings _____

Wainscot: ☐ No   ☐ Yes _____ Thickness _____

DIVISION 9: FINISHES

Barriers: ☐ No ☐ Aluminum ☐ Foil ☐ Mesh ☐ Lead ☐ Leaded Vinyl _____

Acoustical Barriers: ☐ No ☐ Yes _____

Ceilings: ☐ No ☐ Boards ☐ Tile ☐ Cemented ☐ Stapled ☐ On Suspension ☐ _____
  ☐ Fiberglass ☐ Mineral Fiber ☐ Wood Fiber ☐ Metal Pan _____
  ☐ Fire Rated ☐ Ventilating ☐ _____ Ceiling Height _____
  ☐ Luminous Panels _____ ☐ Access Panels _____

Suspension System: ☐ No ☐ T Bar ☐ Z Bar ☐ Carrier Channels ☐ _____

Strip Lighting: ☐ No ☐ Yes _____ Foot Candles _____

Special Acoustical _____
_____

Brick Flooring: ☐ No ☐ Yes _____

Carpet ☐ No ☐ Yes ☐ With Padding ☐ With Backing ☐ _____ Allowance _____
  Type: ☐ Acrylic ☐ Nylon ☐ Polypropylene ☐ Wool ☐ Tile ☐ _____ Face Weight _____
  Padding: ☐ No ☐ Yes _____ Backing: ☐ No ☐ Yes _____

Composition Flooring: ☐ No ☐ Acrylic ☐ Epoxy ☐ Mastic ☐ Neoprene ☐ Polyester ☐ _____
  ☐ Regular Duty ☐ Heavy Duty _____ Thickness _____

Concrete Floor Topping: ☐ No ☐ In Concrete ☐ Yes _____

Resilient Floors: Base: ☐ No ☐ Rubber ☐ Vinyl _____ Height _____
  Asphalt Tile: ☐ No ☐ Yes _____ Color Group _____
  Conductive Tile: ☐ No ☐ Yes _____ Thickness _____
  Cork Tile: ☐ No ☐ Yes _____ Thickness _____
  Linoleum: ☐ No ☐ Yes _____ Thickness _____
  Polyethylene: ☐ No ☐ Yes _____
  Polyurethane: ☐ No ☐ Yes _____ Thickness _____
  Rubber Tile: ☐ No ☐ Yes _____ Thickness _____
  Vinyl: ☐ Sheet ☐ Tile _____ Thickness _____
  Vinyl Asbestos Tile: ☐ No ☐ Yes _____ Color Group _____

Stair Covering: ☐ No ☐ Risers ☐ Treads ☐ Landings ☐ Nosings ☐ Rubber ☐ Vinyl _____

Steel Plates: ☐ No ☐ Cement Bed ☐ Epoxy Bed _____

Wood Floor: ☐ No ☐ Block ☐ Strip ☐ Parquetry ☐ Unfinished ☐ Prefinished ☐ Stock ☐ Custom _____
  Fir: ☐ No ☐ Flat Grain ☐ Vertical Grain _____ Size _____
  Gym: ☐ No ☐ Yes Type _____
  Maple: ☐ No ☐ Yes Grade _____ Size _____
  Oak: ☐ No ☐ Red ☐ White Grade _____ Size _____
  Other: ☐ _____ Grade _____ Size _____
  Finish Required: ☐ No ☐ Yes _____

Wood Block Floor: ☐ No ☐ Creosoted ☐ ☐ Natural _____ Thickness _____

Special Coatings: ☐ No ☐ Floor ☐ Wall _____

Painting: ☐ No ☐ Regular ☐ Fireproof ☐ Fire Retardant ☐ Brush ☐ Roller ☐ Spray _____
  Casework: ☐ No _____ Coats _____ Ceilings: ☐ No _____ Coats _____
  Doors: ☐ No _____ Coats _____ Trim: ☐ No _____ Coats _____
  Walls, Exterior: ☐ No _____ Coats _____ Interior Walls: ☐ No _____ Coats _____
  Windows: ☐ No _____ Coats _____ Piping: ☐ No _____ Coats _____
  Other: _____
  Structural Steel: ☐ No ☐ Yes _____ Miscellaneous Metals: ☐ No ☐ Yes _____

Wall Covering: ☐ No ☐ Cork Tile _____ ☐ Metal Foil _____
  ☐ Flexible Wood Veneers _____ ☐ Vinyl _____ Weight _____
  Wall Paper _____ Vinyl _____ Murals _____
  Other _____

Guards: Corner: ☐ No ☐ Rubber ☐ Steel ☐ Vinyl _____
  Wall: ☐ No ☐ Rubber ☐ Steel ☐ Vinyl _____

SPEC-AID

DIVISION 10: SPECIALTIES

PROJECT _____ LOCATION _____

Bathroom Accessories: ☐ No ☐ Curtain Rod _____ ☐ Dispensers _____ ☐ Grab Bar _____
☐ Hand Dryer _____ ☐ Medicine Cabinet _____ ☐ Mirror _____ ☐ Robe Hook _____
☐ Soap Dispenser _____ ☐ Shelf _____ ☐ Tissue Dispenser _____ ☐ Towel Bar _____
☐ Tumbler Holder _____ ☐ Wall Urn _____ ☐ Waste Receptical _____ ☐ _____
Bulletin Board: ☐ No ☐ Cork ☐ Vinyl Cork ☐ Unbacked ☐ Backed ☐ Stock ☐ Custom
☐ Tan ☐ Framed ☐ No Frames ☐ Changeable Letter ☐ _____ Thickness _____

Canopies: ☐ No ☐ Free Standing ☐ Wall Hung ☐ Stock ☐ Custom _____
Chalkboard: ☐ No ☐ Cement Asbestos ☐ Hardboard ☐ Metal _____ Ga. ☐ Slate _____ in. Thick ☐ Tempered Glass _____
☐ Treated Plastic ☐ _____ ☐ Unbacked ☐ Backed with _____
☐ No Frames ☐ Frames ☐ Chalk Tray ☐ Map Rail ☐ _____
☐ Built in Place ☐ Prefabricated _____
☐ Portable ☐ Reversible ☐ Swing Wing ☐ Sliding Panel _____

Chutes Linen: ☐ No ☐ Aluminum ☐ Aluminized Steel ☐ Stainless Steel ☐ _____ Ga. Diameter _____
☐ Bottom Collector ☐ Sprinklers _____
Mail: ☐ No ☐ Aluminum ☐ Bronze ☐ Stainless ☐ _____ Size _____ ☐ Bottom Collector
Package: ☐ No ☐ Aluminum ☐ Bronze ☐ Stainless _____
Rubbish: ☐ No ☐ Aluminum ☐ Aluminized Steel ☐ Stainless Steel _____ Ga. Diameter _____
☐ Bottom Collector ☐ Sprinklers _____
Compartments & Cubicles: ☐ No ☐ Hospital _____ ☐ Office _____
☐ Shower _____ ☐ Toilet _____ ☐ _____
Control Boards: ☐ No ☐ Yes _____
Decorative Grilles and Screens: ☐ No ☐ Yes _____
Directory Boards: ☐ No ☐ Exterior ☐ Interior ☐ Aluminum ☐ Bronze ☐ Stainless ☐ Lighted _____
Describe _____
Disappearing Stairs: ☐ No ☐ Stock ☐ Custom ☐ Manual ☐ Electric _____ Ceiling Height _____
Display Cases: ☐ No ☐ Economy ☐ Deluxe _____
Fire Extinguishers: ☐ No ☐ CO_2 ☐ Dry Chemical ☐ Foam ☐ Pressure Water ☐ Soda Acid ☐ _____
☐ Aluminum ☐ Copper ☐ Painted Steel ☐ Stainless Steel ☐ _____ Size _____
Cabinets: ☐ No ☐ Aluminum ☐ Painted Steel ☐ Stainless Steel ☐ _____
Hose Equipment: ☐ No ☐ Blanket ☐ Cabinets ☐ Hose _____ Size _____
Protection System: ☐ No ☐ Yes _____
Fireplace, Prefabricated: ☐ No ☐ Economy ☐ Deluxe ☐ Wall Hung ☐ Free Standing _____

Flagpoles: ☐ No ☐ Aluminum ☐ Bronze ☐ Fiberglass ☐ Stainless ☐ Steel ☐ Wood ☐ Tapered ☐ Sectional _____
☐ Ground Set ☐ Wall Set ☐ Counterbalanced ☐ Outriggers _____ Height _____
Bases: ☐ No ☐ Economy ☐ Deluxe _____
Foundation: ☐ No ☐ Yes _____
Folding Gates: ☐ No ☐ Scissors Type ☐ Vertical Members ☐ Stock ☐ Custom _____ Opening _____
Lockers: ☐ No ☐ No Locks ☐ Keyed ☐ Combination _____ Tier Size _____ Height _____
Athletic: ☐ No ☐ Basket ☐ Ventilating ☐ Overhead _____ Size _____
Benches: ☐ No ☐ Yes _____
Special Lockers: _____
Mail Specialties Boxes: ☐ No ☐ Front Loading ☐ Rear Loading ☐ Aluminum ☐ Stainless ☐ _____
Size _____
Letter Slot: ☐ No ☐ Yes _____ Counter Window: ☐ No ☐ Yes _____
Directory: ☐ No ☐ Yes _____ Key Keeper _____
Other: _____

Accordion Folding Partitions: ☐ No ☐ Acoustical ☐ Non Acoustical _____ Weight _____ psf.
Ceiling Height _____ Describe _____

Folding Leaf Partitions: ☐ No ☐ Acoustical ☐ Non Acoustical _____ Weight _____ psf.
Ceiling Height _____ Describe _____

Hospital Partitions: ☐ No ☐ Metal ☐ Curtain Track _____

_____

Movable Office Partitions: ☐ No ☐ Acoustical ☐ Non Acoustical ☐ Asbestos Cement ☐ Hardboard
☐ Laminated Gypsum ☐ Plywood ☐ _____
☐ With Glass ☐ No Glass Describe _____ Partition Height _____
Special Finish: _____
Doors: ☐ No ☐ Yes Type _____ Finish _____ Size _____

Operable Partitions: ☐ No ☐ Yes Type _____

Portable Partitions: ☐ No ☐ Acoustical ☐ Non Acoustical _____ Weight _____ psf.
Partition Height _____ Describe _____

Shower Partitions: ☐ No ☐ Fiberglass ☐ Glass ☐ Marble ☐ Metal ☐ _____ Finish _____
☐ Stock ☐ Custom ☐ Economy ☐ Deluxe Size _____
Doors: ☐ No ☐ Glass ☐ Tempered Glass ☐ Plastic ☐ Curtain Only _____ Size _____
Receptors: ☐ No ☐ Concrete ☐ Metal ☐ Plastic ☐ Terrazzo _____ Size _____
Tub Enclosure: ☐ No ☐ Stock ☐ Custom ☐ Economy ☐ Deluxe _____ Size _____

Toilet Partitions: ☐ No ☐ Fiberglass ☐ Marble ☐ Metal ☐ Slate ☐ Wood ☐ _____
☐ Floor Mounted ☐ Wall Hung ☐ Ceiling Hung _____
Special Finish _____
Doors: ☐ No ☐ Yes _____
Screens: ☐ No ☐ Full Height ☐ Urinal ☐ Floor Mounted ☐ Wall Hung ☐ Ceiling Hung _____

Woven Wire Partitions: ☐ No ☐ Walls ☐ Ceilings ☐ Panel Width _____ Height _____
Doors: ☐ No ☐ Sliding ☐ Swing _____ Windows: ☐ No ☐ Yes
☐ Painted ☐ Galvanized _____

Other Partitions: _____

_____

Parts Bins: ☐ No ☐ Yes _____
Scales: ☐ No ☐ Built in ☐ Portable ☐ Beam Type ☐ Dial Type _____ Capacity _____
Platform Size _____ Material _____ Foundations _____
Accessory Items _____
Shelving, Storage: ☐ No ☐ Metal ☐ Wood _____
Signs: Individual Letters: ☐ No ☐ Aluminum ☐ Bronze ☐ Plastic ☐ Stainless ☐ Steel ☐ _____
☐ Cast ☐ Fabricated Describe _____
Plaques: ☐ No ☐ Aluminum ☐ Bronze _____
Signs: ☐ No ☐ Metal ☐ Plastic ☐ Lighted _____

Sun Control Devices: ☐ No ☐ Yes _____
Telephone Enclosures: ☐ No ☐ Indoor ☐ Outdoor _____
Turnstiles: ☐ No ☐ Yes _____
Vending Machines: ☐ No ☐ Yes _____
Wardrobe Specialties: ☐ No ☐ Yes _____
Other Specialties: _____

_____
_____
_____

SPEC-AID

DATE _____

DIVISION 11: ARCHITECTURAL EQUIPMENT

PROJECT _____ LOCATION _____

Appliances, Residential: ☐ No ☐ Yes Allowance _____ ☐ Separate Contract
☐ Cook Tops _____ ☐ Compactors _____ ☐ Dehumidifier _____ ☐ Dishwasher _____
☐ Dryer _____ ☐ Garbage Disposer _____ ☐ Heaters, Electric _____
☐ Hood _____ ☐ Humidifier _____ ☐ Ice Maker _____ ☐ Oven _____
☐ Refrigerator _____ ☐ Sump Pump _____ ☐ Washing Machine _____ ☐ Water Heater _____
☐ Water Softener _____ ☐ _____

Automotive Equipment: ☐ No ☐ Yes Allowance _____ ☐ Separate Contract
☐ Hoists _____ ☐ Lube _____ ☐ Pumps _____ ☐ _____

Bank Equipment: ☐ No ☐ Yes Allowance _____
☐ Counters _____ ☐ Safes _____ ☐ Vaults _____ ☐ Windows _____
☐ _____

Check Room Equipment: ☐ No ☐ Yes Allowance _____ ☐ Separate Contract
Describe _____

Church Equipment: ☐ No ☐ Yes Allowance _____ ☐ Separate Contract
☐ Altar _____ ☐ Baptistries _____ ☐ Bells & Carillons _____ ☐ Confessionals _____
☐ Organ _____ ☐ Pews _____ ☐ Pulpit _____ ☐ Spires _____
☐ Wall Cross _____ ☐ _____

Commercial Equipment: ☐ No ☐ Yes Allowance _____ ☐ Separate Contract
Describe _____

Darkroom Equipment: ☐ No ☐ Yes Allowance _____ ☐ Separate Contract
Describe _____

Data Processing Equipment: ☐ No ☐ Yes Allowance _____ ☐ Separate Contract
Describe _____

Dental Equipment: ☐ No ☐ Yes Allowance _____ ☐ Separate Contract
☐ Chair _____ ☐ Drill _____ ☐ Lights _____ ☐ X-Ray _____
☐ _____

Dock Equipment: ☐ No ☐ Yes Allowance _____ ☐ Separate Contract
☐ Bumpers _____ ☐ Boards _____ ☐ Door Seal _____ ☐ Levelers _____
☐ Lights _____ ☐ Shelters _____ ☐ _____

Food Service Equipment: ☐ No ☐ Yes Allowance _____ ☐ Separate Contract
☐ Bar Units _____ ☐ Cooking Equip. _____ ☐ Dishwashing Equip. _____ ☐ Food Prep. _____
☐ Food Serving _____ ☐ Refrigerated Cases _____ ☐ Tables _____ ☐ _____

Gymnasium Equipment: ☐ No ☐ Yes Allowance _____ ☐ Separate Contract
☐ Basketball Backstops _____ ☐ Benches _____ ☐ Bleachers _____
☐ Divider Curtain _____ ☐ Gymnastic Equip. _____ ☐ Mats _____ ☐ Scoreboards _____
☐ _____

Industrial Equipment: ☐ No ☐ Yes Allowance _____ ☐ Separate Contract
Describe _____

Laboratory Equipment: ☐ No ☐ Yes Allowance _____ ☐ Separate Contract
☐ Casework _____ ☐ Counter Tops _____ ☐ Hoods _____ ☐ Sinks _____
☐ Tables _____ ☐ _____

Laundry Equipment: ☐ No ☐ Yes Allowance _____ ☐ Separate Contract
☐ Dryers _____ ☐ Washers _____ ☐ _____

Library Equipment: ☐ No ☐ Yes Allowance _____ ☐ Separate Contract
☐ Book Shelves _____ ☐ Book Stacks _____ ☐ Card Files _____ ☐ Carrels _____
☐ Charging Desks _____ ☐ Racks _____ ☐ _____

Medical Equipment: ☐ No ☐ Yes Allowance _____ ☐ Separate Contract
☐ Casework _____ ☐ Exam Room _____ ☐ Incubators _____ ☐ Patient Care _____
☐ Radiology _____ ☐ Sterilizers _____ ☐ Surgery Equip. _____ ☐ Therapy Equip. _____
☐ _____

282

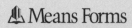
SPEC-AID

DIVISION 11: ARCHITECTURAL EQUIPMENT

Mortuary Equipment: ☐ No ☐ Yes Allowance _____ ☐ Separate Contract
Describe _____

Musical Equipment: ☐ No ☐ Yes Allowance _____ ☐ Separate Contract
Describe _____

Observatory Equipment: ☐ No ☐ Yes Allowance _____ ☐ Separate Contract
Describe _____

Parking Equipment: ☐ No ☐ Yes Allowance _____ ☐ Separate Contract
☐ Automatic Gates _____ ☐ Booths _____ ☐ Control Station _____
☐ Ticket Dispenser _____ ☐ Traffic Detectors _____
☐ _____

Playground Equipment: In Division 2 _____

Prison Equipment: ☐ No ☐ Yes Allowance _____ ☐ Separate Contract
☐ Ceiling Lining _____ ☐ Wall Lining _____ ☐ Bar Walls _____ ☐ Doors _____
☐ Bunks _____ ☐ Lavatory _____ ☐ Water Closet _____ ☐ _____

Residential Equipment: ☐ No ☐ Yes Allowance _____ ☐ Separate Contract
☐ Kitchen Cabinets (Also Div. 6) _____ ☐ Lavatory Cabinets _____ ☐ Kitchen Equipment _____
☐ Laundry Equip. _____ ☐ Unit Kitchens _____ ☐ Vacuum Cleaning _____ ☐ _____
☐ _____

Safes: ☐ No ☐ Yes Allowance _____ ☐ Separate Contract
☐ Office _____ ☐ Money _____ ☐ _____ ☐ Rating _____
Describe _____

Saunas: ☐ No ☐ Yes Allowance _____ ☐ Separate Contract
☐ Built in Place ☐ Prefabricated Size _____ Describe _____
☐ Heater _____ ☐ Seats _____ ☐ Timer _____ ☐ _____

School Equipment: ☐ No ☐ Yes Allowance _____ ☐ Separate Contract
☐ Art & Crafts _____ ☐ Audio-Visual _____ ☐ Language Labs _____ ☐ Vocational _____
☐ Wall Benches _____ ☐ Wall Tables _____ ☐ _____
☐ _____

Shop Equipment: ☐ No ☐ Yes Allowance _____ ☐ Separate Contract
Describe _____

Stage Equipment: ☐ No ☐ Yes Allowance _____ ☐ Separate Contract
Describe _____

Steam Baths: ☐ No ☐ Yes Allowance _____ ☐ Separate Contract
Describe _____

Swimming Pool Equipment: ☐ No ☐ Yes Allowance _____ ☐ Separate Contract
☐ Diving Board _____ ☐ Diving Stand _____ ☐ Life Guard Chair _____ ☐ Ladders _____
☐ Heater _____ ☐ Lights _____ ☐ Pool Cover _____ ☐ Slides _____
☐ _____

Unit Kitchens: ☐ No ☐ Yes Allowance _____ ☐ Separate Contract
Describe _____

Vacuum Cleaning, Central: ☐ No ☐ Yes Allowance _____ ☐ Separate Contract
☐ _____ Valves Describe _____

Waste Disposal Compactors: ☐ No ☐ Yes _____
Incinerators: ☐ No ☐ Electric ☐ Gas Type Waste _____ Capacity _____

Special Equipment: _____
_____
_____
_____
_____
_____

PROJECT _____ LOCATION _____

Artwork: ☐ No ☐ Yes Allowance _____ ☐ Separate Contract
☐ Murals _____ ☐ Paintings _____ ☐ Photomurals _____ ☐ Sculptures _____
☐ Stained Glass _____ ☐ _____

Interior Landscaping: ☐ No ☐ Yes Allowance _____ **Blinds, Exterior:** ☐ No ☐ Yes Allowance _____ ☐ Separate Contract
☐ Solid ☐ Louvered ☐ Aluminum ☐ Nylon ☐ Vinyl ☐ Wood ☐
Describe _____

Blinds, Interior: ☐ No ☐ Yes Allowance _____ ☐ Separate Contract
Folding: ☐ No ☐ Stock ☐ Custom ☐ Wood ☐ _____
Describe _____
Venetian: ☐ No ☐ Stock ☐ Custom ☐ Aluminum ☐ Plastic ☐ Steel ☐ Wood ☐ _____
Describe _____
Vertical: ☐ No ☐ Aluminum ☐ Cloth ☐ Vinyl ☐ _____
Describe _____
Other: _____

Cabinets: ☐ No ☐ Yes Allowance _____ ☐ Separate Contract
☐ Classroom _____
☐ Dormitory _____
☐ Hospital _____
☐ _____

Carpets: In Division 9 _____

Dormitory Units: ☐ No ☐ Yes Allowance _____ ☐ Separate Contract
☐ Beds _____ ☐ Desks _____ ☐ Wardrobes _____ ☐ _____
☐ _____

Drapery & Curtains: ☐ No ☐ Yes Allowance _____ ☐ Separate Contract
Describe _____

Floor Mats: ☐ No ☐ Yes Allowance _____ ☐ Separate Contract
☐ Recessed ☐ Non Recessed _____
☐ Link ☐ Solid _____

Furniture: ☐ No ☐ Yes Allowance _____ ☐ Separate Contract
☐ Beds _____ ☐ Chairs _____ ☐ Chests _____ ☐ Desks _____
Sofas _____ ☐ Tables _____ ☐ _____
☐ _____

Seating Auditorium: ☐ No ☐ Yes Allowance _____ ☐ Separate Contract
Describe _____
Classroom: ☐ No ☐ Yes Allowance _____ ☐ Separate Contract
Describe _____
Stadiuim: ☐ No ☐ Yes Allowance _____ ☐ Separate Contract
Describe _____

Shades: ☐ No ☐ Yes Allowance _____ ☐ Separate Contract
☐ Stock ☐ Custom ☐ Lightproof ☐ Fireproof _____
☐ Cotton ☐ Fiberglass ☐ Vinyl ☐ Woven Aluminum ☐ _____
Describe _____

Wardrobes: ☐ No ☐ Yes Allowance _____ ☐ Separate Contract
☐ Classroom _____ ☐ Dormitory _____ ☐ Hospital _____ ☐ _____
Describe _____

Other Furnishings: _____
_____
_____

PROJECT _____   LOCATION _____

Acoustical Echo Chamber: ☐ No ☐ Yes Allowance _____ ☐ Separate Contract
  Describe _____

  Enclosures: ☐ No ☐ Yes Allowance _____ ☐ Separate Contract
  Describe _____

  Panels: ☐ No ☐ Yes Allowance _____ ☐ Separate Contract
  Describe _____

Air Curtains: ☐ No ☐ Yes Allowance _____ ☐ Separate Contract
  ☐ Heated Air ☐ Unheated Air ☐ Recirculating ☐ Non Recirculating ☐ _____
  Describe _____

Air Inflated Buildings: ☐ No ☐ Yes Describe _____ ☐ Separate Contract

Anechoic Chambers: ☐ No ☐ Yes Allowance _____ ☐ Separate Contract
  Describe _____

Audiometric Rooms: ☐ No ☐ Yes Allowance _____ ☐ Separate Contract
  Describe _____

Bowling Alleys: ☐ No ☐ Yes Allowance _____ ☐ Separate Contract
  Describe _____

Broadcasting Studio: ☐ No ☐ Yes Allowance _____ ☐ Separate Contract
  Describe _____

Chimneys: ☐ No ☐ Yes Allowance _____ ☐ Separate Contract
  Concrete: ☐ No ☐ Unlined ☐ Lined ☐ _____ Diameter _____ Height _____
  Metal: ☐ No ☐ Insulated ☐ Not Insulated ☐ U.L. Listed ☐ Not U.L. Listed ☐ _____
  Describe _____ Diameter _____ Height _____
  Radial Brick: ☐ No ☐ Unlined ☐ Lined ☐ _____ Diameter _____ Height _____
  Foundation: ☐ No ☐ Yes

Clean Rooms: ☐ No ☐ Yes Allowance _____ ☐ Separate Contract
  Describe _____

Comfort Stations: ☐ No ☐ Yes Describe _____

Dark Rooms: ☐ No ☐ Yes Allowance _____ ☐ Separate Contract
  Describe _____

Domes, Observation: ☐ No ☐ Yes Allowance _____ ☐ Separate Contract
  Describe _____

Garage: ☐ No ☐ Yes Describe _____ Cars _____

Garden House: ☐ No ☐ Yes Allowance _____ ☐ Separate Contract
  Describe _____

Grandstand: ☐ No ☐ Yes Describe _____ Seats _____

Greenhouse: ☐ No ☐ Yes Allowance _____ ☐ Separate Contract
  Describe _____

Hangars: ☐ No ☐ Yes Describe _____ Planes _____

Hyperbaric Rooms: ☐ No ☐ Yes Allowance _____ ☐ Separate Contract
  Describe _____

Incinerators (See also Division 10): ☐ No ☐ Yes Allowance _____ ☐ Separate Contract
  Describe _____ Capacity _____

Insulated Rooms: ☐ No ☐ Yes Allowance _____ ☐ Separate Contract
  Doors: ☐ No ☐ Cooler ☐ Freezer ☐ Manual ☐ Electric _____
    ☐ Galvanized ☐ Stainless Describe _____
  Coolers: ☐ No ☐ Yes Describe _____
  Freezers: ☐ No ☐ Yes Describe _____
  Partitions: ☐ No ☐ Yes ☐ Stock ☐ Custom Describe _____
  Other: _____

DIVISION 13: SPECIAL CONSTRUCTION

Integrated Ceilings: ☐ No ☐ Yes Module _____ Ceiling Height _____

  Lighting: ☐ No ☐ Yes Describe _____ Foot Candles _____

  Heating: ☐ No ☐ Yes Describe _____

  Ventilating: ☐ No ☐ Yes Describe _____

  Air Conditioning: ☐ No ☐ Yes Describe _____

Music Practice Rooms: ☐ No ☐ Yes Allowance _____ ☐ Separate Contract

  Describe _____

Pedestal Floors: ☐ No ☐ Yes Allowance _____ ☐ Separate Contract

  ☐ Aluminum ☐ Plywood ☐ Steel ☐ _____ Panel Size _____ Height _____

  ☐ High Density Plastic ☐ Vinyl Tile ☐ V.A. Tile ☐ _____

  Describe _____

Portable Booths: ☐ No ☐ Yes Allowance _____ ☐ Separate Contract

  Describe _____

Prefabricated Structures: ☐ No ☐ Yes Allowance _____ ☐ Separate Contract

  Describe _____

Radiation Protection, Fluoroscopy Room: ☐ No ☐ Yes _____

  Nuclear Reactor: ☐ No ☐ Yes _____

  Radiological Room: ☐ No ☐ Yes _____

  X-Ray Room: ☐ No ☐ Yes _____

  Other: _____

Radio Frequency Shielding: ☐ No ☐ Yes Allowance _____ ☐ Separate Contract

  Describe _____

Radio Tower: ☐ No ☐ Yes Allowance _____ ☐ Separate Contract

  ☐ Guyed ☐ Self Supporting Wind Load _____ psf. _____ Height _____

  Foundations _____

Saunas and Steam Rooms: ☐ No ☐ Yes Allowance _____ ☐ Separate Contract

  Describe _____

Silos: ☐ No ☐ Yes Allowance _____ ☐ Separate Contract

  ☐ Concrete ☐ Steel ☐ Wood ☐ _____ Diameter _____ Height _____

  Foundations _____

Squash & Hand Ball Courts: ☐ No ☐ Yes Allowance _____ ☐ Separate Contract

  Describe _____

Storage Vaults: ☐ No ☐ Yes Allowance _____ ☐ Separate Contract

  Describe _____

Swimming Pool Enclosure: ☐ No ☐ Yes Allowance _____ ☐ Separate Contract

  Describe _____

Swimming Pool Equipment: In Division 11 _____

Swimming Pools: ☐ No ☐ Yes Allowance _____ ☐ Separate Contract

  ☐ Aluminum ☐ Concrete ☐ Gunite ☐ Plywood ☐ Steel ☐ _____

  ☐ Lined ☐ Unlined _____

  Deck: ☐ No ☐ Concrete ☐ Stone _____ Size _____

  Bath Houses: ☐ No ☐ Yes _____ Fixtures _____

Tanks: ☐ No ☐ Yes Allowance _____ ☐ Separate Contract

  ☐ Concrete ☐ Fiberglass ☐ Steel ☐ Wood ☐ _____ Capacity _____

  ☐ Fixed Roof ☐ Floating Roof ☐ _____ Height _____

  Foundations: _____

Therapeutic Pools: ☐ No ☐ Yes Describe _____

Vault Front: ☐ No ☐ Yes Allowance _____ ☐ Separate Contract

  Describe _____ Hour Test _____

Zoo Structures: ☐ No ☐ Yes Describe _____

Other Special Construction: _____

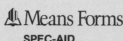 **Means Forms**

SPEC-AID

DATE _____

DIVISION 14: CONVEYING SYSTEMS

PROJECT _____ LOCATION _____

Ash Hoist: ☐ No ☐ Yes Allowance _____ ☐ Separate Contract
Describe _____

Conveyers: ☐ No ☐ Yes Allowance _____ ☐ Separate Contract
Describe _____

Correspondence Lift: ☐ No ☐ Yes Allowance _____ ☐ Separate Contract
Describe _____

Dumbwaiters: ☐ No ☐ Yes Allowance _____ ☐ Separate Contract
Capacity _____ Size _____ Number _____ Floors _____
Stops _____ Speed _____ Finish _____
Describe _____

Elevators, Freight: ☐ No ☐ Yes Allowance _____ ☐ Separate Contract
☐ Hydraulic ☐ Electric ☐ Geared ☐ Gearless _____
Capacity _____ Size _____ Number _____ Floors _____
Stops _____ Speed _____ Finish _____
Machinery Location _____ Door Type _____
Signals _____ Special Requirements _____

Elevators, Passenger: ☐ No ☐ Yes Allowance _____ ☐ Separate Contract
☐ Hydraulic ☐ Electric ☐ Geared ☐ Gearless _____
Capacity _____ Size _____ Number _____ Floors _____
Stops _____ Speed _____ Finish _____
Machinery Location _____ Door Type _____
Signals _____ Special Requirements _____

Escalators: ☐ No ☐ Yes Allowance _____ ☐ Separate Contract
Capacity _____ Size _____ Number _____ Floors _____
Story Height _____ Speed _____ Finish _____
Machinery Location _____ Incline Angle _____
Special Requirements _____

Hoists & Cranes: ☐ No ☐ Yes Allowance _____ ☐ Separate Contract
Describe _____

Lists: ☐ No ☐ Yes Allowance _____ ☐ Separate Contract
Describe _____

Material Handling Systems: ☐ No ☐ Yes Allowance _____ ☐ Separate Contract
☐ Automated ☐ Non Automated ☐ _____
Describe _____

Moving Stairs & Sidewalks ☐ No ☐ Yes Allowance _____ ☐ Separate Contract
Capacity _____ Size _____ Number _____ Floors _____
Story Height _____ Speed _____ Finish _____
Machinery Location _____ Incline Angle _____
Special Requirements _____

Pneumatic Tube System: ☐ No ☐ Yes Allowance _____ ☐ Separate Contract
☐ Automatic ☐ Manual ☐ _____ Size _____ Stations _____
Length _____ Special Requirements _____

Vertical Conveyer: ☐ No ☐ Yes Allowance _____ ☐ Separate Contract
☐ Automatic ☐ Non Automatic ☐ _____
Describe _____

Other Conveying: _____
_____

 **Means Forms**

SPEC-AID

DATE _____

DIVISION 15: MECHANICAL

PROJECT _____   LOCATION _____

Building Drainage: Design Rainfall _____ □ Roof Drains _____ □ Court Drains _____
    □ Floor Drains _____ □ Yard Drains _____ □ Lawn Drains _____ □ Balcony Drains _____
    □ Area Drains _____ □ Sump Drains _____ Shower Drains _____ □ _____
  □ Drain Piping: Size _____ Describe _____
  □ Drain Gates _____ □ Clean Outs _____ □ Grease Traps _____

Sanitary System: □ No   □ Yes   □ Site Main _____ □ Manholes _____
  □ Sump Pumps _____ □ Bilge Pumps _____ □ Ejectors _____
  □ Soils, Stacks _____ □ Wastes, Vents _____ □ _____

Domestic Cold Water: □ No □ Water   Meters _____ □ Law Sprinkler Connection _____
  □ Water Softening _____ □ Water Filtering _____
  □ Boiler Feed Water _____ □ Conditioning Apparatus _____
  □ Standpipe System _____ □ Hose Bibbs _____
  □ Pressure Tank _____ □ Booster Pumps _____
  □ Reducing Valves _____ □ _____

Domestic Hot Water: □ No □ Electric □ Gas □ Oil □ Solar _____
  □ Boiler _____ □ Conditioner _____ □ Fixture Connections _____
  □ Storage Tanks _____ Capacity _____
  □ Pumps _____

Piping: □ No □ Yes   Material _____
  □ Air Chambers _____ □ Escutcheons _____ □ Expansion Joints _____
  □ Shock Absorbers _____ □ Hangers _____
  □ Valves _____ □ Paint _____

Special Piping: □ No □ Compressed Air _____ □ Vacuum _____
  □ Oxygen _____ □ Nitrous Oxygen _____
  □ Carbon Dioxide _____ □ Process Piping _____

Insulation Cold: □ No □ Yes   Material _____ Jacket _____
  Hot: □ No □ Yes   Material _____ Jacket _____

Fixtures Bathtub: □ No □ C.I. □ Steel □ Fiberglass □ _____ Color _____
  □ Curtain □ Rod □ Enclosure □ Wall Shower _____
  Drinking Fountain: □ No □ Yes □ Wall Hung □ Pedestal _____
  Hose Bibb: □ No □ Yes   Describe _____
  Lavatory: □ No □ China □ C.I. □ Steel □ _____ Color _____
    □ Wall Hung □ Legs □ Acid Resisting _____
  Shower: □ No □ Individual □ Group □ Heads □ _____ Size _____
    Compartment: □ No □ Metal □ Stone □ Fiberglass □ _____ □ Door □ Curtain
    Receptor: □ No □ Plastic □ Metal □ Terrazzo □ _____
  Sinks: □ No □ Kitchen _____ □ Janitor _____
  □ Laundry _____ □ Pantry _____
  □ _____
  Urinals: □ No □ Floor Mounted □ Wall Hung _____
    Screens: □ No □ Floor Mounted □ Wall Hung _____
  Wash Centers: □ No □ Yes   Describe _____
  Wash Fountains: □ No □ Floor Mounted □ Wall Hung _____ Size _____
    Describe _____
  Water Closets: □ No □ Floor Mounted □ Wall Hung Color _____
    Describe _____
  Water Coolers: □ No □ Floor Mounted □ Wall Hung _____ Capacity _____ gph.
    □ Water Supply □ Bottle □ Hot □ Compartment _____
  Other Fixtures: _____

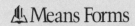 Means Forms

SPEC-AID

DIVISION 15: MECHANICAL

Fire Protection: □ Carbon Dioxide System _____ □ Standpipe _____
□ Sprinkler System   □ Wet   □ Dry _____ Spacing _____
□ Fire Department Connection _____ □ Building Alarm _____
□ Hose Cabinets _____ □ Hose Racks _____
□ Roof Manifold _____ □ Compressed Air Supply _____
□ Hydrants _____ □ _____
Special Plumbing _____
_____

Gas Supply System: □ No   □ Natural Gas   □ Manufactured Gas _____
Pipe: Schedule _____ Fittings _____
Shutoffs: _____ Master Control Valve: _____
Insulation: _____ Paint: _____
Oil Supply System: □ No   □ Tanks   □ Above Ground   □ Below Ground _____
□ Steel   □ Plastic   □ _____ Capacity _____
Heating Plant: □ No   □ Electric   □ Gas   □ Oil   □ Solar _____
□ Boilers _____ □ Pumps _____
□ PRV Stations _____ □ Piping _____
□ Heat Pumps _____
Cooling Plant: □ No   □ Yes _____ Tons _____
Chillers: □ Steam   □ Water   □ Air _____
Condenser—Compressor   □ Air   □ Water _____
Pumps _____ Cooling Towers _____
System Type: _____
□ Single Zone _____ □ Multi-Zone _____
□ All Air _____ □ Terminal Reheat _____
□ Double Duct _____ □ Radiant Panels _____
□ Fan Coil _____ □ Unit Ventilators _____
□ Perimeter Radiation _____ □ _____
Air Handling Units: Area Served _____ Number _____
Total CFM _____ % Outside Air _____
Cooling, Tons _____ Heating, MBH _____
Filtration _____ Supply Fans _____
Economizer _____
Fans: □ No   □ Return   □ Exhaust   □ _____
Describe _____
Distribution: Ductwork _____ Material _____
Terminals: □ Diffusers _____ □ Registers _____
   □ Grilles _____ □ Hoods _____
Volume Dampers: _____
Terminal Boxes: □ High Velocity _____ □ With Coil _____
   □ Double Duct _____ □ _____
Coils: _____
□ Preheat _____ □ Reheat _____
□ Cooling _____ □ _____
Piping: See Previous Page _____
Insulation: Cold: □ No   □ Yes   Material _____ Jacket _____
□ Hot: □ No   □ Yes   Material _____ Jacket _____
Automatic Temperature Controls: _____
Air & Hydronic Balancing: _____
Special HVAC: _____
_____

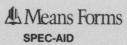 **Means Forms**

SPEC-AID

DATE _____

DIVISION 16: ELECTRICAL

PROJECT _____ LOCATION _____

Incoming Service: □ Overhead   □ Underground

	Primary	Secondary
Voltage _____		
Unit Sub-station & Size _____		
Number of Manholes _____		
Feeder Size _____		
Length _____		
Conduit _____		
Duct _____		
Concrete: □ No □ Yes _____		
Other _____		

Building Service: Size _____ Amps   Switchboard _____

Panels: □ Distribution _____ Lighting _____ Power _____

Describe _____

Motor Control Center: Furnished by _____

Describe _____
_____

Bus Duct: □ No   □ Yes   Size _____ Amps   Application _____

Describe _____

Cable Tray: □ No   □ Yes   Describe _____

Emergency System: □ No   □ Yes   Allowance _____ □ Separate Contract

Generator: □ No   □ Diesel   □ Gas   □ Gasoline _____ Size _____ KW

Transfer Switch: □ No   □ Yes   Number _____ Size _____ Amps

Area Protection Relay Panels: □ No   □ Yes _____

Other _____
_____

Conduit: □ No   □ Yes   □ Aluminum _____

□ Electric Metallic Tubing _____

□ Galvanized Steel _____

□ Plastic _____

Wire: □ No   □ Yes   □ Type Installation _____

□ Armored Cable _____

□ Building Wire _____

□ Metallic Sheath Cable _____

□ _____

Underfloor Duct: □ No   □ Yes   Describe _____
_____

Header Duct: □ No   □ Yes   Describe _____
_____

Trench Duct: □ No   □ Yes   Describe _____
_____

Underground Duct: □ No   □ Yes   Describe _____
_____

Explosion Proof Areas: □ No   □ Yes   Describe _____
_____

Motors: □ No   □ Yes   Total H.P. _____ No. of Fractional H.P. _____ Voltage _____

□ 1/2 to 5 H.P. _____ □ 7-1/2 to 25 H.P. _____ □ Over 25 H.P. _____

Describe _____

Starters: Type _____

Supplied by: _____

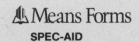

SPEC-AID

DIVISION 16: ELECTRICAL

Telephone System: □ No □ Yes Service Size _____ Length _____
  Manhole: □ No □ Yes Number _____ Termination _____
  Concrete Encased: □ No □ Yes □ Rigid Galv. □ Duct □ _____

Fire Alarm System: □ No □ Yes Service Size _____ Length _____ Wire Type _____
  Concrete Encased: □ No □ Yes □ Rigid Galv. □ Duct □ _____
  □ Stations _____ □ Horns _____ □ Lights _____ □ Combination _____
  Detectors: □ Rate of Rise _____ □ Fixed _____ □ Smoke _____
  Describe _____ Insulation _____ Wire Size _____
  □ Zones _____ □ Conduit _____ □ E.M.T. _____ □ Empty _____
  Describe _____

Watchmans Tour: □ No □ Yes □ Stations _____ □ Door Switches _____
  □ Alarm Bells _____ □ Key Re-sets _____ □ _____
  □ Conduit _____ □ E.M.T. _____ □ Wire _____ □ Empty _____
  Describe _____

Clock System: □ No □ Yes □ Electronic □ Wired □ _____
  □ Single Dial _____ □ Double Dial _____ □ Program Bell _____
  □ Conduit _____ □ E.M.T. _____ □ Empty _____
  Describe _____

Sound System: □ No □ Yes Type _____ Speakers _____
  □ Conduit _____ □ Cable _____ □ E.M.T. _____ □ Empty _____
  Describe _____

Television System: □ No □ Yes Describe _____
  □ Antenna _____ □ Closed Circuit _____ □ Teaching _____ □ Security _____
  □ Learning Laboratory _____ □ _____
  □ Conduit _____ □ E.M.T. _____ □ Wire _____ □ Empty _____

Lightning Protection: □ No □ Yes Describe _____

Low Voltage Switching: □ No □ Yes Describe _____

Scoreboards: □ No □ Yes Describe _____ Number _____

Comfort Systems: □ No □ Electric Heat □ Snow Melting □ _____
  Describe _____

Other Systems: _____
_____

Lighting Fixtures: □ No □ Yes □ Allowance _____ □ Separate Contract
  □ Economy □ Commercial □ Deluxe □ Explosion Proof □ _____
  □ Incandescent _____
  _____ Foot Candles _____
  □ Fluorescent _____
  _____ Foot Candles _____
  □ Mercury Vapor _____
  _____ Foot Candles _____
  □ _____ Foot Candles _____
  □ Step Lighting _____ □ Planter Lighting _____ □ Fountain Lighting _____
  □ Site Lighting _____ □ Poles _____ □ Area Lighting _____ □ Flood Lighting _____
  Dimming System: □ No □ Yes □ Incandescent □ Fluorescent _____
  Ceilings: □ T Bar □ Concealed Spline □ _____
  Emergency Battery Units: □ No □ Lead Acid □ Nickel Cadmium □ 6 Volt _____ 12 Volt _____
  Describe _____

Special Considerations: _____
_____
_____

Index